SpringerBriefs in Fire

Series Editor
James A. Milke

For further volumes:
http://www.springer.com/series/10476

Bryan L. Hoskins • James A. Milke

Study of Movement Speeds Down Stairs

 Springer

Bryan L. Hoskins
Department of Fire Protection
  Engineering
University of Maryland
College Park, Maryland, USA

James A. Milke
Department of Fire Protection
  Engineering
University of Maryland
College Park, Maryland, USA

ISSN 2193-6595   ISSN 2193-6609 (electronic)
ISBN 978-1-4614-3972-1   ISBN 978-1-4614-3973-8 (eBook)
DOI 10.1007/978-1-4614-3973-8
Springer New York Heidelberg Dordrecht London

Library of Congress Control Number: 2012936069

Printed on acid-free paper

Springer is part of Springer Science+Business Media (www.springer.com)

Preface

Building evacuations may be required in the event of a fire, severe weather, or other emergencies. When evacuation is required, the egress system should be designed to enable the people to reach a point of safety before conditions within the building become untenable. In order to design the egress system to meet the expected objectives, the designer needs to be able to accurately predict the time required to evacuate the building.

There are currently dozens of simulation models available for the designer to choose from, but the models are not well validated (Kuligowski et al. 2010). More data and a better understanding of the fundamental principles guiding egress behavior need to be developed (Averill 2011). This makes it even more important to know both what data has been collected to date and the different findings from these studies.

This book focuses on the movement of building occupants on a critical egress component – the stair. The objective is to document the findings of previous literature regarding movement speeds down stairs. This will be accomplished by examining the assumptions, methods, and results of previous studies.

The movement on stairs is critical for determining the amount of time required to safely evacuate the building. For example, the Life Safety Code (NFPA 2009) requires the maximum travel distance to an exit in a new business occupancy that is fully sprinklered to be 91 m. Assuming an approximate travel distance of 8.2 m per floor to a stair (Galbreath 1969), the travel distance on stairs should be greater than the maximum travel distance to reach the stair for buildings taller than 11 stories. Even for buildings less than 11 stories, occupants located closer to the stair may travel further in the stair than outside of it. Furthermore, because the stair shaft is required to be fire-rated and means of egress cannot have occupants move to a less safe location (NFPA 2009), the stair tends to be one of the last components in many complex egress systems. Because the stairs are a major component in the egress system, an error in estimating the movement on stairs will lead to an error in the overall evacuation time.

Previous findings have been used to develop algebraic calculations as well as computer simulation models. The accuracy and applicability of these models are only valid in cases where new data is collected (or estimated) based on the same assumptions as that of the original data. This necessitates understanding what has been done in the previous studies.

The first section of this book describes the different types of studies that have been conducted. The second section details the measurement methods used in these studies for determining the speed and density of people on stairs. The third section then looks at the other variables that have been identified. The final section discusses the implication of the findings. An appendix is then provided that describes the different studies in detail.

References

Averill JD (2011) Five grand challenges in pedestrian and evacuation dynamics. In: Peacock RD, Kuligowski ED, Averill JD (eds) Pedestrian and evacuation dynamics, 2010 Conference, Springer, New York, pp 1–11

Galbreath M (1969) Time of evacuation by stairs in high buildings. National Research Council of Canada, Fire Research Note No. 8

Kuligowski ED, Peacock RD, Hoskins BL (2010) A review of building evacuation models, 2nd edn. National Institute of Standards and Technology, NIST TN-1680

National Fire Protection Association (NFPA) (2009) Life safety code. National Fire Protection Association, Quincy

Contents

List of Figures

List of Tables

Chapter 1
Study Types

In order to develop the algebraic equations or validate the models, data points need to be collected. Data can come from five different types of research. A study can look at behavior of general population on stairs during a fire drill, during normal use, or during a real fire; be a compilation of others' research; or examine a controlled laboratory situation.

1.1 Fire Drills

By observing fire drills, the researcher is able to pre-position recording equipment and staff in order to best observe the evacuation. Drills stress the egress system by requiring large portions of the building to be evacuated at the same time. Building occupants can interpret a fire drill as an actual emergency (Khristy 1985), so the behavior is similar to what would be seen during an actual fire for occupants remote from the effects of the fire. Examples of researchers that have observed fire drills include Pauls (1980) and Proulx (1995).

1.2 Normal Use

Observing normal use conditions is similar to observing fire drills with respect to the ease of observation. Unlike during drills, there is no indication of an emergency. These studies have typically been conducted in locations where people are expected to move with some urgency. Examples include transit stations (e.g., London Transport Board 1958) and theaters (e.g., National Bureau of Standards 1935). Fruin (1971) and Templer (1975) also collected normal use data.

B.L. Hoskins and J.A. Milke, *Study of Movement Speeds Down Stairs*,
SpringerBriefs in Fire, DOI 10.1007/978-1-4614-3973-8_1, © The Author(s) 2013

1.3 Real Fires

Unlike data collected during fire drills or normal use, data collected during real fires do not require any assumptions about behavior being similar to real events. Unfortunately, there is no known data publicly available that systematically records movement on stairs during a real fire. Instead, researchers have interviewed survivors after the fact. There are two main limitations to this approach. First, it assumes that a person's memory is perfect (or can be adjusted by the researcher to match known facts). Second, it is naturally biased to only include those individuals that survived the fire and are willing to discuss their experiences. Examples of this type of study include Galea and Blake (2004) and Averill et al. (2005).

1.4 Compiled Works

Compiled studies collect the works of others and attempt to find similarities and general guidance from multiple sources. They do not collect any data for their analysis. Thus, in theory, they eliminate the bias that might be present in a particular study. However they are ultimately only as valid as the sources they used. Both Predtechenskii and Milinskii (1978) and Gwynne and Rosenbaum (2008) are examples of compiled works.

1.5 Laboratory Studies

Laboratory studies are conducted under controlled conditions. The researcher is interested in specific details and the subjects are instructed on exactly how to behave; there is no sense of emergency or other "real" effect. Examples of this type of study include Templer (1975), Frantzich (1996), and Boyce et al. (1999).

References

Averill JD, Mileti DS, Peacock RD, Kuligowski ED, Groner N, Proulx G, Reneke PA, Nelson HE (2005) Occupant behavior, egress, and emergency communication. Federal building and fire safety investigation of the World Trade Center disaster. National Institute of Standards and Technology, NIST NCSTAR 1-7

Boyce KE, Shields TJ, Silcock GWH (1999) Toward the characterization of building occupancies for fire safety engineering: capabilities of disabled people moving horizontally and on an incline. Fire Technol 35(1):51–67

Frantzich H (1996) Study of movement on stairs during evacuation using video analysis techniques. Department of Fire Safety Engineering, Lund Institute of Technology, Lund University

Fruin JJ (1971) Pedestrian planning and design. Metropolitan Association of Urban Designers and Environmental Planners, New York

Galea ER, Blake S (2004) Collection and analysis of human behaviour data appearing in the mass media relating to the evacuation of the World Trade Centre towers of 11 September 2001. Office of the Deputy Prime Minister

Gwynne SMV, Rosenbaum ER (2008) Employing the hydraulic model in assessing emergency movement. In: DiNenno P (ed) The SFPE handbook of fire protection engineering, 4th edn. National Fire Protection Association, Quincy

Khisty CJ (1985) Pedestrian flow characteristics on stairways during disaster evacuation. Transp Res Rec 1047:97–102

London Transport Board (1958) Second report of the operational research team on the capacity of footways. London Transport Board research report No. 95

National Bureau of Standards (1935) Design and construction of building exits. National Bureau of Standards, miscellaneous publication M151

Pauls JL (1980) Building evacuation: research findings and recommendations. In: Cantor D (ed) Fires and human behaviour. Wiley, New York, pp 251–275

Predtechenskii VM, Milinskii AI (1978) Planning for foot traffic flow in buildings (trans: Sivaramakrishnan MM). Amerind Publishing, New Delhi

Proulx G (1995) Evacuation time and movement in apartment buildings. Fire Saf J 24(3): 229–246

Templer JA (1975) Stair shape and human movement. PhD dissertation, Columbia University

Chapter 2
Measurement Methods

Several authors have previously conducted studies that looked at movement speeds and human behavior while descending stairs. However, when estimating the movement speed of people, these authors tended to concentrate on the general characteristics of the entire population rather than those of the individuals. The previous studies tended to only consider average values for all observations or individuals in isolation.

Detailed descriptions of the studies are available in Appendix A. This section focuses on the speeds that the different authors observed and the key variables that they indicated play a role in movement speeds down stairs. For each of the studies presented, as much data as was possible from the authors' descriptions was given with respect to the physical conditions and interactions of the building and stairs as well as the occupants. Unless otherwise noted, all comments that give explanations for effects are from the authors of the studies.

2.1 Movement Speed Measurements

Speed was determined in the different studies by how long it took individuals to travel a known distance. While the time component of the movement speed is relatively consistent, the distance measurement is not. When the authors described how they calculated it, several different methods were used. In some instances (e.g., Kagawa et al. 1985; Shields et al. 2009) times were given on a per floor basis. However most of the authors instead calculated speed based on a measured distance within the stairs.

When descending stairs in a high-rise building, travel distance in stairs needs to include both travel on landings and treads. The differences in the studies regarding travel distance stemmed from which components were included and, if included, how the travel distance was calculated.

B.L. Hoskins and J.A. Milke, *Study of Movement Speeds Down Stairs*,
SpringerBriefs in Fire, DOI 10.1007/978-1-4614-3973-8_2, © The Author(s) 2013

2.1.1 Landing Calculation

Some authors (e.g., London Transport Board 1958; Frantzich 1994) chose to ignore travel on landings. These studies involved only occupants while they were on the treads. Other authors (e.g., Predtechenskii and Milinskii 1978; Peacock et al. 2011) chose to calculate travel distance by assuming that the travel distance on the landing would be twice the stair width. Hoskins (2011) used a landing length of $\pi/2$ times the stair width. Unfortunately, many other authors (e.g., Pauls and Jones 1980; Galea et al. 2009) did not indicate whether landings were included or not or, if included, how the distance was calculated. The known method used in each study is shown in Table 2.1.

2.1.2 Treads Calculation

All of the studies included occupants traveling along the treads. As with the landings, two different methods were used. First, (e.g., Pauls 1980; Daly et al. 1991) travel distance was along the slope of the stairs. Second, (e.g., Fruin 1971a; Templer 1992) travel distance was only calculated in the horizontal direction. In many instances, (e.g., Khisty 1985; Proulx 1995) the authors did not indicate which method they used. The known method used for each study is also shown in Table 2.1.

2.1.3 Combined Travel Distance Calculation Method

Of the 44 references involving movement speeds on stairs, only 23 reported how travel distance was calculated for at least one component. The methods used by the different authors are shown in Table 2.1. The remaining 21 references did not indicate how travel distance was calculated on either the landings or treads and thus are not included in the table.

2.2 Observed Movement Speeds

With nearly half of the references not providing details about how the travel distance was calculated, direct comparisons between different studies need to be done with caution. As can be seen in Table 2.2, there is a range of movement speeds that have been observed or predicted by different authors (with known horizontal speeds converted to slope speeds based on stair dimensions).

Table 2.1 Travel distance calculation methods

Study type	Study	Landing method	Treads method
Compiled	Predtechenskii and Melinskii (1978)	Linear path	Slope
Real fires	Galea and Blake (2004)	Linear path	Slope
Fire drills	Kratchman (2007)	Linear path	Slope
Fire drills	Blair (2010)	Linear path	Slope
Fire drills	Peacock et al. (2011)	Linear path	Slope
Fire drills	Hoskins (2011)	Arc path	Slope
Compiled	Galbreath (1969)	Linear path	Horizontal
Laboratory	Frantzich (1994)	Not considered	Slope
Fire drills	Kagawa et al. (1985)	Not considered	Not considered
Real fires	Shields et al. (2009)	Not considered	Not considered
Normal use	London Transport Board (1958)	Not considered	Not provided
Fire drills	Pauls (1971)	Not provided	Slope
Fire drills	Pauls and Jones (1980)	Not provided	Slope
Fire drills	Pauls (1980)	Not provided	Slope
Normal use	Daly et al. (1991)	Not provided	Slope
Laboratory	Frantzich (1996)	Not provided	Slope
Laboratory	Boyce et al. (1999)	Not provided	Slope
Compiled	Gwynne and Rosenbaum (2008)	Not provided	Slope
Normal use	Ye et al. (2008)	Not provided	Slope
Real fires	Galea et al. (2009)	Not provided	Slope
Normal use	Fruin (1971a)	Not provided	Horizontal
Compiled	Templer (1992)	Not provided	Horizontal
Laboratory	Fujiyama and Tyler (2004)	Not provided	Horizontal

The average values from Table 2.2 are shown in Fig. 2.1. In the instances where speeds were given on a per floor basis, the distance per floor was assumed to be 8 m (Galbreath 1969). For the horizontal speed, the tread depth and riser height were assumed to be 0.2794 and 0.1778 m respectively (Pauls 1984).

There is no one value that authors consistently report for the average speed. While Melinek and Booth (1975), Pauls (1980), and Proulx (2008) recommended using 0.5 m/s, none of the studies where data was actually collected[1] reported averages (or ranges of averages) that fell completely with 20% of this recommended value. Individual data points will lie even further away from this value than averages do.

2.3 Density Measurements

Many of the authors have determined relationships between movement speed and density to better predict movement speeds. As was the case with travel distance, authors have used a variety of methods to calculate density.

---

[1] Excluding the compiled studies that only gave rules of thumb or generalized findings.

Table 2.2 Observed movement speeds

Study type	Study	Average speed	Minimums or maximums
Fire drill	Pauls (1971)	0.61–0.81 m/s	
Fire drill	Pauls and Jones (1980)	0.44–0.66 m/s	0.23 m/s (min)
Fire drill	Pauls (1980)	0.50 m/s	
Fire drill	Khisty (1985)		0.64 m/s (norm)
			0.70 m/s (drill)
Fire drill	Kagawa et al. (1985)	16 s/flr	
Fire drill	Proulx (1995)	0.52–0.62 m/s	
Fire drill	Proulx et al. (1995)	0.95–1.07 m/s	
Fire drill	Proulx et al. (1996)	0.75–1.2 m/s	
Fire drill	Shields et al. (1997)	0.33–1.1 m/s	
Fire drill	Proulx et al. (1999)		0.39–1.30 m/s
Fire drill	Kratchman (2007)	0.70–0.80 m/s	
Fire drill	Proulx et al. (2007)	0.40–0.66 m/s	0.17–1.87 m/s
Fire drill	Hostikka et al. (2007)	0.64 m/s	0.5–1.5 m/s
Fire drill	Peacock et al. (2009)	0.40–0.83 m/s	
Fire drill	Blair (2010)	0.37–0.57 m/s	0.01–1.43 m/s
Fire drill	Peacock et al. (2011)	0.48 m/s	0.056–1.7 m/s
Fire drill	Hoskins (2011)	0.53–0.61 m/s	0.07–2.02 m/s
Normal use	NBS (1935)	0.45–0.65 m/s	
Normal use	London Transport Board (1958)	0.67–0.98 m/s	
Normal use	Fruin (1971a)	0.56–1.10 m/s	
Normal use	Daly et al. (1991)	0.56–0.67 m/s	
Normal use	Tanaboriboon and Guyano (1991)	0.58–0.62 m/s	0.39–0.89 m/s
Normal use	Lee and Lam (2006)	0.48–0.65 m/s	0.29–0.93 m/s
Normal use	Ye et al. (2008)		0.5–1.2 m/s
Real fires	Galea and Blake (2004)	0.2–0.7 m/s	
Real fires	Averill et al. (2005)	0.2 m/s	
Real fires	Shields et al. (2009)	43–150 s/flr	
Real fires	Galea et al. (2009)	0.29 m/s	
Compiled	Melinek and Booth (1975)	0.5 m/s	
Compiled	Predtechenskii and Milinskii (1978)	0.18–0.27 m/s	
Compiled	Templer (1992)	0.45 m/s (horiz)	
Compiled	Smith (1995)		0.1–0.9 m/s
Compiled	Proulx (2008)	0.5 m/s	0.76[a] m/s
Laboratory	Frantzich (1994)	1.0 m/s	0.3–1.3 m/s
Laboratory	Frantzich (1996)	0.69–0.72 m/s	2.27 m/s (max)
Laboratory	Boyce et al. (1999)	0.13–0.70 m/s	0.11–1.10 m/s
Laboratory	Wright et al. (2001)	0.30–0.42 m/s	
Laboratory	Fujiyama and Tyler (2004)	0.60–1.30 m/s	

[a] Indicates value is an average

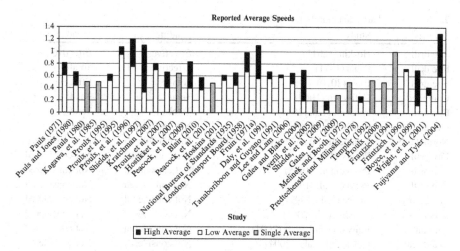

Fig. 2.1 Average speeds in references

Some studies (e.g., Proulx 1995; Shields et al. 2009) ignored density entirely. Predtechenskii and Milinskii (1978) calculated density based on the fraction of floor space occupied by individuals. For the other authors that were interested in density (e.g., Galbreath 1969; Gwynne and Rosenbaum 2008), density was calculated based on the number of persons per unit area of floor space. The area used was either on the treads or a combination of treads and landings, yet another variation across studies. The methods used for each component were not consistent to complicate matters even more.

2.3.1 Fluctuations in Density

Most authors did not indicate how they handled whether density was static or dynamic. With three exceptions average values (whether for the entire evacuation or over short durations) were assigned to all occupants. Blair (2010) assumed that the density for occupants was a variable for each individual, and could be calculated based on the number of people in the entire area over multiple floors. Peacock et al. (2011) also assumed that density changed with each individual, but the density within a single camera view held constant over multiple floors. Hoskins (2011) used the average value on adjacent cameras for each individual.

Table 2.3 Total area calculation methods

Study type	Study	Landing area	Tread area
Fire drill	Hoskins (2011)	Effective	Effective
Fire drill	Pauls (1971)	Effective	Total
Fire drill	Blair (2010)	Total	Effective
Compiled	Galbreath (1969)	Total	Total
Fire drill	Peacock et al. (2011)	Total	Total
Fire drill	Kratchman (2007)	Not considered	Effective
Normal use	Tanaboriboon and Guyano (1991)	Not considered	Not considered
Fire drill	Proulx (1995)	Not considered	Not considered
Fire drill	Proulx et al. (1995)	Not considered	Not considered
Fire drill	Proulx et al. (1996)	Not considered	Not considered
Laboratory	Boyce et al. (1999)	Not considered	Not considered
Laboratory	Wright et al. (2001)	Not considered	Not considered
Laboratory	Fujiyama and Tyler (2004)	Not considered	Not considered
Real fires	Galea and Blake (2004)	Not considered	Not considered
Real fires	Averill et al. (2005)	Not considered	Not considered
Normal use	Lee and Lam (2006)	Not considered	Not considered
Real fires	Shields et al. (2009)	Not considered	Not considered
Fire drill	Proulx et al. (2007)	Not considered	Not provided
Fire drill	Pauls (1980)	Not provided	Effective
Normal use	Daly et al. (1991)	Not provided	Effective
Compiled	Gwynne and Rosenbaum (2008)	Not provided	Effective
Compiled	Joint Committee (1952)	Not provided	Total
Normal use	London Transport Board (1958)	Not provided	Total

2.3.2   Landing Area

Very few authors provided any information about how they calculated landing areas. Some authors (e.g., Kratchman 2007; Proulx et al. 2007) chose to only use treads. Pauls (1971) used an effective area to account for occupants not normally being in the corners, but no other details were provided. Blair (2010) calculated the area based on the total area of the landings. The methods of authors that indicated how landing areas were calculated are shown in Table 2.3.

2.3.3   Tread Area

As with the landings, many authors did not state how the area of treads was calculated. While some may have chosen to not include treads, with one exception, the other authors that described their method accounted for the horizontal surface of the treads. Kratchman (2007) used the area of the slope of the treads. There was also a difference between whether an effective area should be used (e.g., Pauls 1980; Daly et al. 1991), accounting for a boundary layer people left to the edge of the stair, or the total width (e.g., Joint Committee 1952; London Transport Board 1958). When known, the method used in each study is provided in Table 2.3.

2.3.4 Combined Landing and Tread Areas

Only 23 of the 44 references provided information about how density was calculated. And, within this subset, the information provided by half of the studies was that they did not measure it. Only five authors that included landings gave any indication as to how it was calculated. Eleven authors (including the five that gave information about landing areas) included information in their studies that indicated how they measured areas on the treads.

2.4 Observed Densities

As shown in Table 2.3, a majority of the authors did not clearly indicate how density was measured. Table 2.4 shows the density values reported by the different authors.

The range of densities reported in the different references are shown in Fig. 2.2 with average or optimum values shown where the authors did not provide the complete range of values.

As was the case with velocities, there was little agreement between the authors as to what characteristic densities are. This was evident in the reported average values as well as in the minimum, maximum, and optimum values.

2.5 Equations

Despite differences in the reported movement speeds and densities, many of the authors have provided data or proposed equations to predict movement speeds based on density. These equations are shown in Table 2.5.

All of the equations for movement speed based on density are shown in Fig. 2.3.

Even for densities where the different equations are the closest to converging, there is still a range of predicted speeds that is greater than 0.3 m/s. Other variables not accounted for in the formulas could explain the wide variation in proposed equations.

Table 2.4 Observed densities

Study type	Study	Average density	Reported minimums or maximums
Fire drill	Pauls and Jones (1980)	1.38 persons/m^2	
Fire drill	Pauls (1980)		2.0 persons/m^2 (optimum)
Fire drill	Khisty (1985)	1.38 persons/m^2 (normal) 1.40 persons/m^2 (emergency)	
Fire drill	Kagawa et al. (1985)		<3.0 persons/m^2
Fire drill	Proulx et al. (1999)	1.00–2.05 persons/m^2	
Fire drill	Proulx et al. (2007)	1.56–1.60 persons/m^2	2.30 persons/m^2
Fire drill	Hostikka et al. (2007)		0.5–2.5 persons/m^2
Fire drill	Blair (2010)	0.886–1.329 persons/m^2	0.019–3.653 persons/m^2
Fire drill	Hoskins (2011)	1.01–1.73 persons/m^2	0.28–3.51 persons/m^2
Normal use	National Bureau of Standards (1935)	1.3–2.6 persons/m^2	2.8 persons/m^2 (maximum)
Normal use	London Transport Board (1958)		1.6 persons/m^2 (optimum)
Normal use	Fruin (1971a)	0.72–1.08 persons/m^2	0.54 persons/m^2 (end of free-flow) 2.70 persons/m^2 (maximum)
Compiled	Melinek and Booth (1975)	2.2 persons/m^2	
Compiled	Predtechenskii and Milinskii (1978)		0.01–0.92 m^2/m^2
Compiled	Smith (1995)		<4.0 persons/m^2
Compiled	Proulx (2008)		0.54–3.2 persons/m^2
Compiled	Gwynne and Rosenbaum (2008)		0.54–3.8 persons/m^2

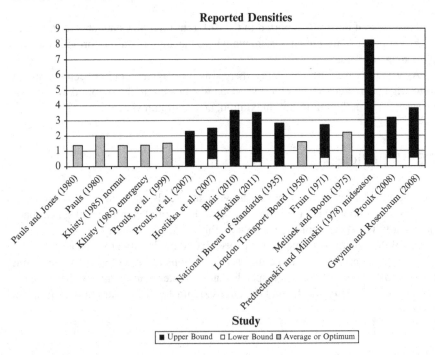

Fig. 2.2 Reported densities in references

Table 2.5 Movement speed formulas

Study type	Study	Formula	
Fire drill	Pauls (1980)	$s = 1.08 - 0.29 \cdot D$	(2.1)
Fire drill	Khisty (1985)	$s = 0.864 - 0.187 \cdot D$ (normal)	(2.2)
		$s = 0.798 - 0.177 \cdot D$ (emergency)	(2.3)
Fire drill	Shields et al. (1997)	$s = 1.27 - 0.30 \cdot D$ (ahead of wheelchair)	(2.4)
		$s = 1.69 - 0.05 \cdot D$ (behind wheelchair text)	(2.5)
		$s = 1.69 - 0.50 \cdot D$ (behind wheelchair graph)	(2.6)
Fire drill	Hoskins (2011)	$s = \dfrac{1}{1.93 + 0.51 \cdot D}$ (from slowest occupants)	(2.7)
		$s = \dfrac{1}{0.30 + 0.35 \cdot D}$ (upper bound)	(2.8)
		$s = \dfrac{1}{1.93 + 1.67 \cdot D}$ (lower bound)	(2.9)
Normal use	Fruin (1971b)	$s = 0.650 - 0.097 \cdot D$	(2.10)
Normal use	Daly et al. (1991)	$s = \dfrac{1}{1.5 + 0.3 \cdot \left(f / 1.14 \right)^{2.7}}$ [a]	(2.11)
Normal use	Ye et al. (2008)	$s = 0.996 - 0.159 \cdot D$ [a]	(2.12)

(continued)

Table 2.5 (continued)

Study type	Study	Formula	
Compiled	Predtechenskii and Milinskii (1978)	$s = gathered\left(112 \cdot d^4 - 380 \cdot d^3 + 434 \cdot d^2 - 217 \cdot d + 57\right)$ $\left(0.775 + 0.44 \cdot e^{(-0.39 \cdot d)} \cdot \sin\left(5.61 \cdot d - 0.224\right)\right) gathered / 60$	(2.13)
		$s = 1.0308 \cdot x^2 - 1.7867 \cdot x + 0.861$ (normal)[a]	(2.14)
		$s = 0.0159 \cdot x^2 - 0.2443 \cdot x + 1.0418$ (emergency)[a]	(2.15)
Compiled	Pauls (1984)	$s = 1.08 - 0.29\left\lvert \dfrac{1}{\left((b-\delta)/P\right)y} \right\rvert$	(2.16)
Compiled	Smith (1995)	$s = 0.9 - 0.13 \cdot D$ [a]	(2.17)
Compiled	Gwynne and Rosenbaum (2008)	$s = k - 0.266 \cdot k \cdot D$	(2.18)

[a] Portions of the equations extrapolated from other information in the text

where:

s = speed (m/s)

D = density (persons/m²)

d = density (m²/m²)

f = specific flow (persons/m-s)

k = 1.00 (m/s) for 19.0 cm riser, 25.4 cm tread

= 1.08 (m/s) for 17.8 cm riser, 27.9 cm tread

= 1.16 (m/s) for 16.5 cm riser, 30.5 cm tread

= 1.23 (m/s) for 16.5 cm riser, 33.0 cm tread

b = stair width (m)

δ = boundary layer (m)

P = population (persons)

y = depth (m)

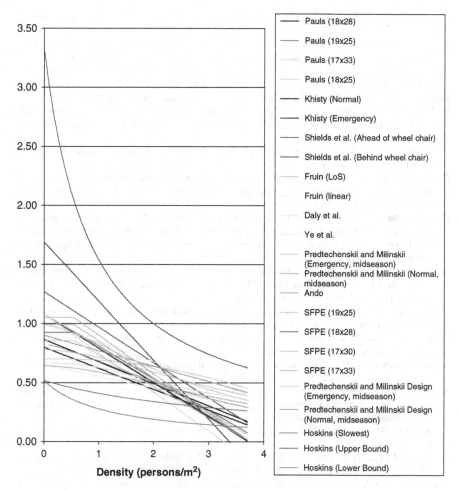

Fig. 2.3 Movement speed equations based on density

References

Averill JD, Mileti DS, Peacock RD, Kuligowski ED, Groner N, Proulx G, Reneke PA Nelson HE (2005) Occupant behavior, egress, and emergency communication. Federal building and fire safety investigation of the World Trade Center disaster. National Institute of Standards and Technology, NIST NCSTAR 1–7

Blair AJ (2010) The effect of stair width on occupant speed and flow of high rise buildings. MS thesis, University of Maryland, College Park

Boyce KE, Shields TJ, Silcock GWH (1999) Toward the characterization of building occupancies for fire safety engineering: capabilities of disabled people moving horizontally and on an incline. Fire Technol 35(1):51–67

Daly PN, McGrath F, Annesley TJ (1991) Pedestrian speed/flow relationships for underground stations. Traffic Eng Control 32(2):75–78

Frantzich H (1994) A model for performance-based design of escape routes. Department of Fire Safety Engineering, Lund Institute of Technology, Lund University

Frantzich H (1996) Study of movement on stairs during evacuation using video analysis techniques. Department of Fire Safety Engineering, Lund Institute of Technology, Lund University

Fruin JJ (1971a) Pedestrian planning and design. Metropolitan Association of Urban Designers and Environmental Planners, New York

Fruin JJ (1971b) Designing for pedestrians: a level of service concept. Transp Res Rec 355:1–15

Fujiyama T, Tyler N (2004) An explicit study on walking speeds of pedestrians on stairs. Presented at 10th international conference on mobility and transport for elderly and disabled people, Hamamatsu

Galbreath M (1969) Time of evacuation by stairs in high buildings. National Research Council of Canada, fire research note No. 8

Galea ER, Blake S (2004) Collection and analysis of human behaviour data appearing in the mass media relating to the evacuation of the World Trade Centre towers of 11 September 2001. Office of the Deputy Prime Minister

Galea ER, Hulse L, Day R, Siddiqui A, Sharp G (2009) The UK WTC 9/11 evacuation study: an overview of the methodologies employed and some analysis relating to fatigue, stair travel speeds and occupant response times. In: Proceedings of the 4th international symposium on human behaviour in fire, Robinson College, Cambridge, pp 27–40

Gwynne SMV, Rosenbaum ER (2008) Employing the hydraulic model in assessing emergency movement. In: DiNenno P (ed) The SFPE handbook of fire protection engineering, 4th edn. National Fire Protection Association, Quincy

Hoskins BL (2011) The effects of interactions and individual characteristics on egress down stairs. PhD dissertation, University of Maryland, College Park

Hostikka S, Paloposki T, Rinne T, Saari J, Korhonen T, Heliövaara S (2007) Evacuation experiments in offices and public buildings. VTT, Working Papers 85

Joint Committee (1952) Fire grading of buildings part III precautions relating to personal safety. Post-war Building Studies Number 29:22–95

Kagawa M, Kose S, Morishita Y (1985) Movement of people on stairs during fire evacuation Drill-Japanese experience in a Highrise office building. Fire safety science In: Proceedings of the 1st international symposium, Gaithersburg, pp 533–540

Khisty CJ (1985) Pedestrian flow characteristics on stairways during disaster evacuation. Transp Res Rec 1047:97–102

Kratchman JA (2007) An investigation on the effects of firefighter counterflow and human behavior in a six-story building evacuation. MS thesis, University of Maryland, College Park

Lee JYS, Lam WHK (2006) Variation of walking speeds on a unidirectional walkway and on a bidirectional stairway. Transp Res Rec 1982:122–131

London Transport Board (1958) Second report of the operational research team on the capacity of footways. London Transport Board research report No. 95

Melinek SJ, Booth S (1975) An analysis of evacuation times and the movement of crowds in buildings. Building research establishment, current paper 96/75

National Bureau of Standards (1935) Design and construction of building exits. National Bureau of Standards, miscellaneous publication M151

Pauls JL (1971) Evacuation drill held in the BC hydro building 26 June 1969. National Research Council of Canada, building research report 80

Pauls JL (1980) Building evacuation: research findings and recommendations. In: Cantor D (ed) Fires and human behaviour. Wiley, New York, pp 251–275

Pauls J (1984) The movement of people in buildings and design solutions for means of egress. Fire Technol 20(1):27–47

Pauls JL, Jones BK (1980) Building evacuation: research methods and case studies. In: Cantor D (ed) Fires and human behaviour. Wiley, New York, pp 227–249

Peacock RD, Averill JD, Kuligowski ED (2009) Stairwell evacuations from buildings: what we know we don't know. National Institute of Standards and Technology, NIST technical note 1624

Peacock RD, Hoskins BL, Kuligowski ED (2011) Overall and local movement speeds during fire drill evacuations in buildings up to 31 stories. In: Peacock RD, Kuligowski ED, Averill JD (eds) Pedestrian and evacuation dynamics, 2010 conference. Springer, New York, pp 25–36

Predtechenskii VM, Milinskii AI (1978) Planning for foot traffic flow in buildings (trans: Sivaramakrishnan MM). Amerind Publishing, New Delhi

Proulx G (1995) Evacuation time and movement in apartment buildings. Fire Saf J 24(3):229–246

Proulx G (2008) Evacuation timing. In: DiNenno P (ed) The SFPE handbook of fire protection engineering, 4th edn. National Fire Protection Association, Quincy

Proulx G, Latour JC, Maclaurin JW, Pineau J, Hoffman LE, Laroche C (1995) Housing evacuation of mixed ability occupants in highrise buildings. National Research Council of Canada, internal report 706

Proulx G, Kaufman A, Pineau J (1996) Evacuation times and movement in office buildings. National Research Council of Canada, internal report 711

Proulx G, Tiller DK, Kyle BR, Creak J (1999) Assessment of photoluminescent material during office occupant evacuation. National Research Council of Canada, internal report 774

Proulx G, Bénichou N, Hum JK, Restivo KN (2007) Evaluation of the effectiveness of different photoluminescent stairwell installations for the evacuation of office building occupants. National Research Council of Canada, research report 232

Shields TJ, Boyce KE, Silcock GWH, Dunne B (1997) The impact of a wheelchair bound evacuee on the speed and flow of evacuees in a stairway during an uncontrolled unannounced evacuation. J Appl Fire Sci 7(1):29–39

Shields TJ, Boyce KE, McConnell N (2009) The behaviour and evacuation experiences of WTC 9/11 evacuees with self-designated mobility impairments. Fire Saf J 44:881–893

Smith RA (1995) Density, velocity, and flow relationships for closely packed crowds. Saf Sci 18:321–327

Tanaboriboon Y, Guyano JA (1991) Analysis of pedestrian movements in Bangkok. Transp Res Rec 1294:52–56

Templer JA (1992) The staircase: studies of hazards, falls and safer design. The MIT Press, Cambridge

Wright MS, Cook GK, Webber GMB (2001) The effects of smoke on people's walking speeds using overhead lighting and Wayguidance provision. Human behavior in fire, In: Proceedings of the 2nd international symposium, MIT

Ye J, Chen X, Yang C, Wu J (2008) Walking behavior and pedestrian flow characteristics for different types of walking facilities. Transp Res Rec 2048:43–51

Chapter 3
Additional Variables

In some of the referenced studies, authors reported other variables that were present. In some instances, the authors stated that these other variables could explain some of the variation in the data. All of the variables that the different authors identified are discussed in the following section.

3.1 Perception of Drill

The behavior of occupants under drill conditions could be different than how they would behave under real fire emergencies. If differences exist, this could impact movement speeds.

Early researchers believed that there was a difference in movement speeds based on a sense of urgency. The National Bureau of Standards (1935) found that occupants at a transit station moved faster than those at a theater. However other variables were not controlled. The Joint Committee (1952) stated that occupants would behave in an urgent manner during an actual emergency, but this was not quantified. Predtechenskii and Milinskii (1978) assumed that movement speeds near a fire would be greater than those further away (where speeds would be similar to normal movement speeds).

Later researchers conducted surveys of drill participants to determine what perceptions those people had as well as how the individuals that thought it was a drill behaved compared to those that thought it was an actual emergency. Khisty (1985), in exit interviews with a random sample of at least 10% of the occupants after each drill, found that 80% of occupants thought that the drills were real incidents. Thus, for those individuals, their behavior under drill conditions would be the same as under a real emergency with remote cues. Proulx et al. (1996) surveyed participants in two fire drills and asked them about their perceptions of the drill as well as actions that they undertook. In the first drill, 21.2% believed that it was an actual emergency and, in the second drill, 23.2% believed it was an actual emergency. When comparing

B.L. Hoskins and J.A. Milke, *Study of Movement Speeds Down Stairs*,
SpringerBriefs in Fire, DOI 10.1007/978-1-4614-3973-8_3, © The Author(s) 2013

reported actions between the occupants that believed it was not a drill and those that did, there were no statistical differences in actions taken in the first building. In the second building, the occupants that thought it was an actual emergency were statistically more likely to save files and turn off their computers.

3.2   Stair Width

Two different authors (Khisty 1985; Frantzich 1996) found that the width of the stairs did not alter movement speeds. Khisty (1985) also stated that localized constrictions similarly did not alter movement speeds, but no description was given to define what was meant by a localized constriction.

3.3   Stair Conditions

There were three types of conditions within the stairs that were identified by authors as causing changes in movement speed. The first were environmental factors in real fire events that slowed down occupants or forced them to choose an alternate route. The second type of condition relates to changing densities. Finally, the riser height and tread depth were found to alter movement speeds.

Galea and Blake (2004) and Shields et al. (2009) reported that water in the stairs hindered movement speed. Furthermore, Shields et al. (2009) found that strong odors made people leave the stair.

During some evacuations (e.g., Pauls and Jones 1980; Kagawa et al. 1985; Blair 2010), densities increased at intermediary floors and decreased at lower floors. The authors did not provide explanations as to what caused the density to increase on certain floors and not on others. These increases in density corresponded to locations at which the movement speeds decreased.

As for movement speeds changing due to tread dimensions, Predtechenskii and Milinskii (1978) stated that speeds on very steep stairs were less than those on more gradual inclines. Pauls (1984) theorized that movement speeds would change (within certain limits) depending on the riser height and tread depth. These assumptions were used to develop the equations presented by Gwynne and Rosenbaum (2008).

3.4   Exit Selection

Typically, occupants were found to use a well-known egress path (Pauls 1980; Proulx 1995; Proulx et al. 1996; Shields et al. 1997). While Proulx et al. (1995) reported occupants using the nearest exit, they did not indicate if there was a difference in familiarity between the exits. In some instances this led to an unequal usage of the different stairs as closer stairs were bypassed. The more commonly used stairs then had slower movement speeds than the lighter used stairs.

3.5 Weather

In most studies, the effects of weather were not considered. Two studies did make observations about the effects of weather on movement speeds. In the first, during cold or wet weather, occupants wearing coats were found to have a decrease in flow of approximately 6% (Pauls 1980). On the other hand, one drill observed by Proulx et al. (1995) was conducted during winter weather and the speeds were slower, but not significantly slower, than those drills conducted during warmer weather.

3.6 Groups

In some studies, groups formed and changed composition as occupants descended (e.g., Galea and Blake 2004). Galea and Blake (2004) found that most (88% and 90%) of the occupants they studied from the World Trade Center evacuation traveled in groups. Other authors also identified groups being present in their studies.

Two different types of groups were identified in the different studies. The first were groups that formed due to either the building evacuation plan or by occupants staying with previous acquaintances (e.g., Pauls 1971; Kagawa et al. 1985). In residential settings, occupants were found to evacuate in small groups with large spaces between groups (Proulx 1995; Proulx et al. 1995). The second types of groups formed spontaneously as occupants descended. Both types can be seen in a single evacuation (e.g., Shields et al. 2009). Kratchman 2007 indicated groups being present, but did not indicate how they formed.

Proulx et al. (1995) reported that occupants traveling in groups moved slower than individuals by themselves (1.00 compared to 1.18 m/s) and that the speed of the groups was at the speed of their slowest member. Unfortunately, the authors did not explain how they reached this conclusion that the speed was of the slowest member rather than the group dynamics causing the slowest people to go faster than they usually would, but still slower than the average person.

Potentially related to groups, Hoskins (2011) identified flow units within the larger flow and followers within the flow. It could not be determined from this data if the flow units were groups of occupants choosing to descend together or just clusters of people moving at the same rate.

3.7 Occupant Spacing

As occupants descended the stairs, some authors indicated patterns in how they were spaced. Pauls (1971) and Kagawa et al. (1985) found that occupants tended to walk down in a staggered file or two abreast with an open tread between pairs. Thus, the minimum spacing was one person per tread when two exit lanes were available. Similarly, Fruin (1971) found that occupants tended to be four to five treads apart during normal movement and that the maximum density (where the flow came to a stop) was when there was one person on every other tread. Hoskins (2011) used the number of open treads between occupants rather than persons/m^2 or m^2/m^2. Proulx et al. (2007) observed occupants at higher densities moving in a staggered file as had Pauls (1971) and Kagawa et al. (1985). At lighter densities, Proulx et al. (2007) found that occupants tended to stay to the right. The staying to one side phenomenon was also found by Galea and Blake (2004) for able-bodied occupants in the World Trade Center and by Hoskins (2011) for slower moving occupants during heavy density conditions. During light density conditions, Hoskins (2011) found that occupants tended to stay to the inside of the stair. One study (Peacock et al. 2011) used a multiple regression model and found that which side of the stair occupants were on was significant in determining their movement speed.

One author attempted to quantify the line-of-travel distance between occupants. Frantzich (1996) found that the minimum interpersonal spacing was usually 0.37 m (free-flow speed was possible at this distance), but, when instructed to be as close to the person ahead as possible, the minimum spacing was 0.25 m.

Aside from the spacing between occupants, how they were spaced on a tread was also observed by some authors. Occupants were found to leave a boundary layer, but to also stay near the handrails while descending stairs (Pauls 1980). Previous work (e.g., London Transport 1958; Joint Committee 1952) had included the entire width of the steps, in terms of exit lanes, when calculating the density.

3.8 Gender

Studies were divided as to whether or not movement speeds were significantly different depending on gender. Proulx et al. (1995) found gender to not be significant in movement speed in two of three buildings, but males were statistically faster in the third building. In only one building in one study (Proulx et al. 1996) were females found to be statistically faster, but, for the other building in the study, males had a greater, but not significantly so, average speed. Kratchman (2007) found that males moved faster, but the difference was not significant. Hostikka et al. (2007) and Peacock et al. (2009) found that there was no difference in movement speed based on gender.

On the other hand, Fruin (1971) found that males moved faster than females. And Peacock et al. (2011) found that males moved statistically faster than females when other variables were controlled for. Hoskins (2011) found that gender was significant for individuals in light density conditions, but not significant in heavy density conditions.

In some instances, gender was also found to influence occupants' behavior. Pauls and Jones (1980) found that there was a "ladies first" deference as occupants entered from a given floor. Proulx e al. (1996) found that females in one building required less pre-evacuation time. Kratchman (2007) and Hoskins (2011) found that females were nearly three times more likely to be carrying objects than were males. Hoskins

(2011) also found that same gender pairings were more common than opposite gender pairings and that the pairings with male followers were descending at a faster rate than the comparable pairings with female followers.

3.9 Age

The exact age ranges that authors used varied from one study to another. In some instances (e.g., Proulx et al. 1996) speeds were not found to be dependent on age. There were other authors that did find differences in movement speed based on age.

Proulx (1995) found that very young and very old occupants tended to move slower than the other occupants and Proulx et al. (1995) found that seniors moved significantly slower than other age groups. Similarly, Fruin (1971) found that speed decreased for older occupants.

Fujiyama and Tyler (2004) reported that older and younger subjects, when asked to descend at a normal pace, did not move at statistically different speeds. However, when the subjects were asked to descend as quickly as possible, the younger subjects went statistically faster.

Proulx (1995) also found differences in behavior based on age. Specifically, older individuals tended to begin their evacuation sooner. However Proulx et al. (1995) found that seniors were not significantly different than the rest of the population in two buildings and required more time in the third.

3.10 Carrying

Three studies attempted to quantify the change in movement speed based on whether the occupants were carrying items or not. Proulx (1995) and Hoskins (2011) found that occupants carrying items tended to move at the same speed as other occupants. However Proulx (1995) noted that they tended to behave in a more cautious manner. Peacock et al. (2011) used a multiple regression model and found that occupants carrying items went significantly slower than those that were not when the other variables were held constant.

3.11 Handrail Use

Using handrails, in at least two studies, was found to impact how occupants used the stairs. Proulx et al. (2007) observed 70–90% of occupants on the upper floors using handrails (the percentage decreased lower in the building). The use of handrails caused occupants to progress in single-file. Boyce et al. (1999) found that 94% of unassisted disabled subjects used the handrail.

In one study that attempted to quantify how handrail use alters movement speed, Peacock et al. (2011) found no difference in movement speeds between occupants that used or did not use handrails.

3.12   Fatigue

Some authors suggested that fatigue was present (e.g., Shields et al. 1997) and others suggested that it was not (e.g., Khisty 1985). However none of the authors explained how they reached this conclusion.

In other instances, fatigue was determined based on general observations. Proulx et al. (1999) stated that occupants that exited below the main floors of their study moved faster in part due to less fatigue, but this was not quantified. Galea and Blake (2004) reported some instances of fatigue and that this was typically caused by footwear. Galea et al. (2009) reported occupants stopping when their companions needed to stop due to fatigue.

Quantification of how fatigue would slow down occupants was based on both assumptions by the authors and by statistical analysis. The Joint Committee (1952) believed that occupants would slow down by 8% for every 3.05 m (above 6.10 m) if occupants did not have to slow down due to merging flows. Similar to this assumption, Hoskins (2011) found that occupants in light density conditions slowed as they descended while those in heavy density conditions did not. Peacock et al. (2009) found that travel distance was significant when they used a multiple regression model to predict movement speeds. On the other hand, Peacock et al. (2011) found that travel distance was not significant in predicting movement speed.

3.13   Body Size

Most studies did not consider the effect of body size on movement speed. The three that did, (Fujiyama and Tyler 2004; Galea et al. 2009; Hoskins 2011) found no differences in speed based on body mass index or body size.

3.14   Pre-evacuation Time

No studies have looked directly at pre-evacuation time and movement speeds on stairs. Five studies, (Proulx 1995; Proulx et al. 1995, 1996, 2007; Shields et al. 1997) reported pre-evacuation times that varied from 0.6 to 9.7 min depending on the building being studied. Proulx (1995), Proulx et al. (1995, 1996), and Shields et al. (1997) measured the time after the alarm to when occupants left their apartment, office, or room. The three articles written by Proulx referred to this time as the "time to start". Proulx et al. (2007) observed when occupants entered the stairs and assumed that time to start or pre-evacuation time (the authors stating that the two terms were equivalent) would be 10–15 s less than this value.

None of the studies made a connection between pre-evacuation time and movement speed. However Proulx et al. (1996) found that women had shorter times than men in the building where women moved faster, but the authors did not provide the decreased pre-evacuation time as a possible reason for the greater speeds.

In their regression model for predicting movement speeds, Peacock et al. (2009) found that pre-evacuation time was significant. However, in the regression model by Peacock et al. (2011), pre-evacuation time was not significant. Hoskins (2011) found that the effects of the time the occupants were first seen in the stairs was not consistent across buildings. In all three instances, pre-evacuation time was measured based on when occupants entered the stairs.

3.15 Passing

The studies disagreed as to how and whether passing would take place. Pauls (1980) found that occupants would pass slower or disabled occupants as they descended. He claimed that these occupants did not alter the overall flow because occupants, once past the slower moving occupant, were able to fill any gaps in the flow that had been created. However Shields et al. (1997) found that occupants were unwilling to pass a wheelchair user being assisted down the stairs despite there being approximately 40 cm to do so and Proulx et al. (2007) found that occupants using the handrail or with disabled occupants ahead of them were not passing slower moving occupants.

If occupants did pass, there was further disagreement as to whether it was the slower individuals being passed or the faster individuals passing that changed their path. Hoskins (2011) found that occupants engaging in passing behavior were more likely to be in the inner lane and those individuals being passed were more likely to be in the outer lane. Shields et al. (2009) found that some interviewees had engaged in passing behavior while others had allowed others to pass them. In one instance, a group of people formed behind a slower moving occupant and chose not to pass. Kratchman (2007) observed occupants engaging in passing behavior, with faster moving individuals moving to the outside. Lee and Lam (2006) observed some individuals weaving through the crowd and passing other individuals. On the other hand, Hostikka et al. (2007) observed that passing behavior occurred when slower moving individuals moved to the side to allow others to pass them. Galea and Blake (2004) reported able-bodied occupants staying to one side to allow injured occupants to pass using the other side of the stair. Galea et al. (2009) reported occupants stopping to allow others to pass them. Other studies (i.e., Frantzich 1996; Ye et al. 2008) observed some occupants engaging in passing behavior, but did not indicate how the dynamics occurred.

3.16 Merging

Three different types of merging are possible. In the first, occupants in the stairs defer to occupants entering the stairs (Pauls and Jones 1980; Proulx et al. 1996; Shields et al. 2009). In the second, occupants on the floors defer to those already in the stair (Hostikka et al. 2007). The third type is where neither defers and the occupants on the floor and already in the stair split evenly. Kagawa et al. (1985) had occupants report (from different floors) that there were instances where the occupants in the stairs would not let occupants from the floors enter and that there were instances where occupants entering from the floor caused severe disruptions to the flow in the stairs. As occupants merged into the stair, the flow would slow down or become stagnant (Proulx et al. 2007). Hoskins (2011) found that individuals that allowed other people to enter the flow moved slower than most other occupants during that portion of the evacuation.

3.17 Counterflow

Some early research (London Transport Board 1958) indicated that counterflow did not have an impact on movement speed. More recent studies (e.g., Galea and Blake 2004; Kratchman 2007; Lee and Lam 2006; Peacock et al. 2009) have found the opposite to be true. Furthermore, Daly et al. (1991), when adjusting flows based on factors developed for level surfaces, found that the flows on the stairs were still less than expected when counterflow was present. As for the dynamics of counterflow, Kratchman (2007) noted that, to accommodate the counterflow, occupants were observed to move to the right.

References

Blair AJ (2010) The effect of stair width on occupant speed and flow of high rise buildings. MS thesis, University of Maryland, College Park

Boyce KE, Shields TJ, Silcock GWH (1999) Toward the characterization of building occupancies for fire safety engineering: capabilities of disabled people moving horizontally and on an incline. Fire Technol 35(1):51–67

Daly PN, McGrath F, Annesley TJ (1991) Pedestrian speed/flow relationships for underground stations. Traffic Eng Control 32(2):75–78

Frantzich H (1996) Study of movement on stairs during evacuation using video analysis techniques. Department of Fire Safety Engineering, Lund Institute of Technology, Lund University

Fruin JJ (1971) Pedestrian planning and design. Metropolitan Association of Urban Designers and Environmental Planners, New York

Fujiyama T, Tyler N (2004) An explicit study on walking speeds of pedestrians on stairs. Presented at 10th international conference on mobility and transport for elderly and disabled people, Hamamatsu

Galea ER, Blake S (2004) Collection and analysis of human behaviour data appearing in the mass media relating to the evacuation of the World Trade Centre towers of 11 September 2001. Office of the Deputy Prime Minister

Galea ER, Hulse L, Day R, Siddiqui A, Sharp G, (2009) The UK WTC 9/11 evacuation study: an overview of the methodologies employed and some analysis relating to fatigue, stair travel speeds and occupant response times. In: Proceedings of the 4th international symposium on human behaviour in fire, Robinson College, Cambridge, pp 27–40

Gwynne SMV, Rosenbaum ER (2008) Employing the hydraulic model in assessing emergency movement. In: DiNenno P (ed) The SFPE handbook of fire protection engineering, 4th edn. National Fire Protection Association, Quincy

Hoskins BL (2011) The effects of interactions and individual characteristics on egress down stairs. PhD dissertation, University of Maryland, College Park

Hostikka S, Paloposki T, Rinne T, Saari J, Korhonen T, Heliövaara S (2007) Evacuation experiments in offices and public buildings. VTT, working papers 85

Joint Committee (1952) Fire grading of buildings part III precautions relating to personal safety. Post-war Building Studies Number 29:22–95

Kagawa M, Kose S, Morishita Y (1985) Movement of people on stairs during fire evacuation drill-Japanese experience in a highrise office building. Fire safety science. In: Proceedings of the 1st international symposium, Gaithersburg, pp 533–540

Khisty CJ (1985) Pedestrian flow characteristics on stairways during disaster evacuation. Transp Res Rec 1047:97–102

Kratchman JA (2007) An investigation on the effects of firefighter counterflow and human behavior in a six-story building evacuation. MS thesis, University of Maryland, College Park

Lee JYS, Lam WHK (2006) Variation of walking speeds on a unidirectional walkway and on a bidirectional stairway. Transp Res Rec 1982:122–131

London Transport Board (1958) Second report of the operational research team on the capacity of footways. London Transport Board research report No. 95

National Bureau of Standards (1935) Design and construction of building exits. National Bureau of Standards, miscellaneous publication M151

Pauls JL (1971) Evacuation drill held in the BC hydro building 26 June 1969. National Research Council of Canada, building research report 80

Pauls JL (1980) Building evacuation: research findings and recommendations. In: Cantor D (ed) Fires and human behaviour. Wiley, New York, pp 251–275

Pauls J (1984) The movement of people in buildings and design solutions for means of egress. Fire Technol 20(1):27–47

Pauls JL, Jones BK (1980) Building evacuation: research methods and case studies. In: Cantor D (ed) Fires and human behaviour. Wiley, New York, pp 227–249

Peacock RD, Averill JD, Kuligowski ED (2009) Stairwell evacuations from buildings: what we know we don't know. National Institute of Standards and Technology, NIST technical note 1624

Peacock RD, Hoskins BL, Kuligowski ED (2011) Overall and local movement speeds during fire drill evacuations in buildings up to 31 stories. In: Peacock RD, Kuligowski ED, Averill JD (eds) Pedestrian and evacuation dynamics, 2010 conference. Springer, New York, pp 25–36

Predtechenskii VM, Milinskii AI (1978) Planning for foot traffic flow in buildings (trans: Sivaramakrishnan MM). Amerind Publishing, New Delhi

Proulx G (1995) Evacuation time and movement in apartment buildings. Fire Saf J 24(3):229–246

Proulx G, Latour JC, Maclaurin JW, Pineau J, Hoffman LE, Laroche C (1995) Housing evacuation of mixed ability occupants in highrise buildings. National Research Council of Canada, internal report 706

Proulx G, Kaufman A, Pineau J (1996) Evacuation times and movement in office buildings. National Research Council of Canada, internal report 711

Proulx G, Tiller DK, Kyle BR, Creak J (1999) Assessment of photoluminescent material during office occupant evacuation. National Research Council of Canada, internal report 774

Proulx G, Bénichou N, Hum JK, Restivo KN (2007) Evaluation of the effectiveness of different photoluminescent stairwell installations for the evacuation of office building occupants. National Research Council of Canada, research report 232

Shields TJ, Boyce KE, Silcock GWH, Dunne B (1997) The impact of a wheelchair bound evacuee on the speed and flow of evacuees in a stairway during an uncontrolled unannounced evacuation. J Appl Fire Sci 7(1):29–39

Shields TJ, Boyce KE, McConnell N (2009) The behaviour and evacuation experiences of WTC 9/11 evacuees with self-designated mobility impairments. Fire Saf J 44:881–893

Ye J, Chen X, Yang C, Wu J (2008) Walking behavior and pedestrian flow characteristics for different types of walking facilities. Transp Res Rec 2048:43–51

Chapter 4
Summary

All of the studies previously presented made observations about movement speeds down stairs. The methods used in calculating these speeds were not consistent and the values varied by large amounts. In an effort to explain the disparity between different observations, several authors found that movement speed was dependent on density. Once again, the method for calculating this variable was not consistent. Multiple equations were developed where speed decreased with density, but the values predicted by the different equations for specific densities are dissimilar. Some authors then attempted to examine if other variables were responsible for the differences in movement speeds.

For algebraic equations, the density equations have been commonly used, but multiple methods have been used to define how the variables were measured. These differences in measurement methods could explain part of the variation between the different algebraic equations that have been proposed. An approach similar to one of the methods proposed by Hoskins and Milke (2012) will allow for a direct comparison of different studies.

For the sophisticated computer models, other variables can be taken into account. As shown by Hoskins (2011), these other variables can lead to a much better predictive method than simply relying on density alone. Seventeen of these potential variables have been described in this brief. Further research can be used to refine these variables and understand the interactions between them.

With the stairs being one of the last egress components, any errors in the calculations involving them will negate the accuracy of calculations from upstream components. As shown in Chap. 2, a wide range of equations has been proposed. By including consideration of variables like those in Chap. 3, the accuracy of the predictions can be increased.

B.L. Hoskins and J.A. Milke, *Study of Movement Speeds Down Stairs*,
SpringerBriefs in Fire, DOI 10.1007/978-1-4614-3973-8_4, © The Author(s) 2013

References

Hoskins BL (2011) The effects of interactions and individual characteristics on egress down stairs. PhD dissertation, University of Maryland, College Park

Hoskins BL, Milke JA (2012) Differences in measurement methods for travel distance and area for estimates of occupant speed on stairs. Fire Saf J 48:49–57

Appendix A
Details of Previous Studies

This appendix contains details of the studies mentioned in the main portion of the brief. These articles were selected because they provided an equation, rule of thumb, or observed data of people descending stairs. Fire drills, normal use, actual fires, compiled woks, and laboratory studies are all included.

For each of the studies presented, as much data as was possible from the authors' descriptions was given with respect to the physical conditions and interactions of the building and stairs as well as the occupants. Unless otherwise noted, all comments that give explanations for effects are from the authors of the studies.

Fire Drill Studies

When an emergency evacuation is required, the egress system needs to be able to accommodate the people in the building at that time. No known systematic observations of movement speeds have been made under actual fire conditions. For ethical reasons, intentional fire experiments are not possible. Thus, for systematic analysis that can be preplanned, drill data has to be used.

It is unknown how accurate drill data for movement speeds is compared to speeds in actual emergencies. The only indication comes from indirect sources. In real fires, interviews with victims after the fact indicate that people, once remote from the fire, tend to behave normally and in an altruistic manner (e.g. Keating 1982). Also, the initial reactions of people not receiving fire cues can have significant delays in starting the egress process (e.g., Chertkoff and Kushigian 1999; Kuligowski and Hoskins 2011). These are actions that are typically associated with fire drills.

B.L. Hoskins and J.A. Milke, *Study of Movement Speeds Down Stairs*,
SpringerBriefs in Fire, DOI 10.1007/978-1-4614-3973-8, © The Author(s) 2013

Studies have been identified where the authors observed actual fire drills and were able to observe the evacuation speeds. These measurements were made in some instances by having observers move with the occupants and in other instances by using videotape evidence.

Pauls – BC Hydro Building

Pauls (1971) observed the evacuation of the BC Hydro Building on June 26, 1969. The drill was pre-announced and 910 occupants used two 1.19 m stairs with 17.8 cm riser heights and 25.4 cm tread depths. An additional 35 occupants that could not walk down the stairs were able to use elevators for egress. Occupants were originally located on floors 1–21 over a total of 17,500 m² of floor space. Occupants on a given floor reported to the exit and waited there until being instructed to enter the stair; the protocol was for the lower floors to be evacuated first. Occupants tended to evacuate in groups with the spacing being two abreast or in a staggered file. People moving within the main flow made observations. For a few selected instances, the author was able to estimate the average descent speed and these values ranged from 0.61 to 0.81 m/s. No information was provided about the basis for these calculations. The average discharge rate was 0.7 persons/s with a peak flow of 1.2 persons/s.

For calculating the density (for the flow calculation) an adjusted horizontal area was used to account for people not using the corners of the landings; the entire width of the stairs was used. Aside from the two flow values, the author did not give any indication of the density. Travel distance was measured long the slope of the stair, but the path on the landings was not defined.

Pauls and Jones

Pauls and Jones (1980) studied two different office buildings. The comparison between the two unannounced evacuations focused on total evacuation versus phased evacuation. Both buildings were medium-sized, high-rise, government office buildings located in Ottawa, Canada. Conversely, the plan configuration was markedly different and the building used for total evacuation had nearly four times the effective stair width.

The building used for the total evacuation had 32,500 m² of office space over 14 floors. There were five 1.14 m-wide dogleg stairs. The riser height and tread depth were not provided. The drill was conducted on a cool October 1972 day with 1,453 able-bodied people using four of the five stairs and 73 people that were disabled or assisting the disabled occupants using the center stair. Seventeen observers collected data with five at ground level and one moving person every five stories for each of the four stairs used by able-bodied occupants. The observers had tape recorders to provide data on human behavior, densities, and movement speeds.

In this study, occupants already in the stairs deferred to people entering from lower floors. Above the seventh floor, the movement speeds, measured along the slope of the stairs, varied due to the stairs being used at a significantly greater density. The slowest recorded speed was 0.23 m/s. Below the seventh floor, the average speed was 0.44 m/s. On the last floor, the mean speed was 0.66 m/s with an average density of 1.38 persons/m².

The building used for the phased evacuation had 30,800 m² of office space over 20 floors. There were only two dogleg exit stairs that were 1.04 m wide. The drill was conducted in May 1971. The phased evacuation was designed to evacuate the fire floor first, the two adjacent floors second, and then the other floors starting from the top of the building. Observers moved with other occupants on floors 3, 4, 12, 14, and 21 and others were positioned on the ground and second floors in fixed positions. The observers followed the same procedures as the total evacuation drill.

During the drill, unclear instructions over the public address system caused several floors to evacuate out of sequence. While the authors did not provide movement speeds, the observers during steady-state conditions descended at approximately the same rate as those in the total evacuation study under steady-state conditions.

Pauls – Multiple Buildings

From the late 1960s to the 1970s, Pauls (1980) conducted 58 total evacuations from high-rise office buildings. The stairs had widths that ranged from 0.91 to 1.52 m. The variation of riser heights and tread dimensions were not provided, but mention was made of the maximum tread depth, 27.9 cm, and at least one stair with a tread depth of 22.9 cm. Also, the exact heights of buildings were not given, but buildings 18–20 stories were described as being very tall. Typically, people stayed near the sides of the stairs in a staggered file, but left a space between themselves and the wall.

Based on these observations, the author developed a formula to calculated the expected flow on a stair:

$$F = 0.206 \cdot (b - 0.3) \cdot \left(\frac{P}{b - 0.3} \right)^{0.27} \qquad (A.1)$$

where:

F = flow (persons/s)
P = population (persons)
b = width of stairs (m)

In Eq. A.1, the −0.3 term is to account for the space (a boundary layer) that people left between themselves and the wall.

Pauls noted that individual characteristics could alter the total time required for evacuation. He identified 20 of the 58 cases as being ones where individuals required coats due to cold or wet weather. In these evacuations, the flows dropped by 6%.

Also, individuals tended to more frequently use stairs for evacuations that they used under normal conditions and those buildings with more training tended to have higher flows. He also found that disabled occupants slowed the flow in their general vicinity, but had no noticeable effect on the overall flow.

In order to calculate total evacuation times, he proposed using:

$$t = 2.00 + 0.117 \cdot p \qquad (A.2)$$
$$t = 0.70 + 0.133 \cdot p \qquad (A.3)$$

where:

t = time (min.)
p = evacuation population per meter of effective width

Equation A.2 was to be used when p < 800 persons/m of effective width and Eq. A.3 was to be used when p > 800 persons/m of effective width. Pauls also found that movement speed was dependent on density:

$$s = 1.08 - 0.29 \cdot D \qquad (A.4)$$

where:

s = speed (m/s)
D = density (persons/m^2)

The speed was the speed along the slope of the stair. To calculate the horizontal component, the author said to multiply the speed by 0.9. There was also no indication as to how the travel distances on the landings were calculated.

Furthermore, the author concluded that the optimum evacuation conditions occurred when the density was 2.0 persons/m^2 and the speed was 0.5 m/s. However he did not explain how density was to be calculated. A figure in the chapter showed an overhead camera shot of occupants only on treads. If that figure is representative of the data that he collected, then the density was for the treads only, but it is not stated whether this was the case or not.

While not stated, Eq. A.4 makes several assumptions as to the nature of speed on stairs. One is that the relationship between speed and density is a linear relationship. The equation also implies that an individual in isolation will travel at 1.08 m/s and that, at a density of 3.7 people/m^2, all movement will stop.

For calculating the flow, Pauls suggested using:

$$f = 1.26 \cdot D - 0.33 \cdot D^2 \qquad (A.5)$$

where:

f = specific flow (persons/m-s)
D = density (persons/m^2)

If the theory relating specific flow, speed, and density was completely accurate without any other effects, then Eq. A.5 would be Eq. A.4 multiplied by the density. Thus, at least for the data collected by Pauls, there appears to possibly be some nonlinear interaction between flow, speed, and density.

Khisty

Khisty (1985) observed 21 unannounced fire drills and normal use in dormitories 3–12 stories in height on the Washington State University, Pullman campus during 1983 and 1984. Drills were conducted at all hours of the day with the latest one at 11 p.m. Exit interviews with a random sample of at least 10% of the occupants after each drill found that 80% of occupants thought that the drills were real incidents. Thus, for those individuals, their behavior under drill conditions would be the same as under a real emergency with remote cues.

In this study, scissor stairs were most common, occurring in 19 of the 21 buildings. The risers varied from 16.5 to 19.0 cm with tread heights from 27.9 to 30.5 cm. The width of the stairs varied from 1.22 to 2.13 m. As expected for a dormitory, about 99% of the occupants were between 18 and 30 years old. Time-lapse photography at 18 frames per second was used to record the drills. Observers also moved within the flow to collect data.

Movement speed was calculated based on the number of frames between two marked locations a known distance apart. The author did not indicate if the speed was calculated for the slope or horizontal component. The density was calculated based on the number of individuals within the area between the two marked locations, but the author did not indicate if that was an effective area or if it was of just the treads, landings, or a combination of the two. The flows were calculated by multiplying the density by the speed.

During the emergency evacuation, the mode and median of density were 1.96 and 1.40 persons/m² respectively. Comparatively, under normal conditions, the mode and median of density were 1.66 and 1.38 persons/m² respectively. The highest recorded speeds were 0.635 and 0.696 m/s for normal and emergency conditions respectively. Flows were also seen to increase under emergency conditions. The maximum specific flow increased from 0.898 to 0.998 persons/m-s.

Equations were provided for movement speeds down stairs for both normal and emergency conditions.

$$s = 0.864 - 0.187 \cdot D \qquad (A.6)$$
$$s = 0.798 - 0.177 \cdot D \qquad (A.7)$$

where:

s = speed (m/s)
D = density (persons/m²)

Equation A.6 is for normal conditions and Eq. A.7 is for emergency conditions.

Several variables did not seem to decrease the speeds or flows. These included the stair width and localized constrictions; a description of the constrictions was not provided. Also, fatigue was not observed, but how this was determined was not stated.

Kagawa, Kose, and Morishita

Kagawa et al. (1985) recorded a fire drill in a 53 story high-rise office building in Tokyo, Japan on September 4, 1984 at 14:30. Approximately 1,500 individuals, 20% of the building population, participated in the drill and used the two 1.20 m-wide emergency stairs. The occupants had been notified in advance about the drill.

Four pairs of video cameras were used to monitor the flow through selected doors on the east stair. An unspecified number of additional cameras were used to monitor the general flow. Research staff with cameras also moved with the last person on selected floors.

The first individuals exited the building 42 s after the initial alarm. Most of the occupants were outside of the building within 16 min. The flow was not uniform but consisted of groups moving as platoons and occupants were either in a staggered file or should-to-shoulder with an open tread to the preceding person. Generally speaking, the observers had an initial delay after first entering the stairs and then descended approximately one story every 16 s. Each story was approximately 3.65 m high. While not provided in the article, based on these numbers, the approximate vertical travel speed was 0.23 m/s. No values were given for the horizontal or slope components.

Stagnation of the flow was reported in several locations with people on some of the lower floors commenting in questionnaires that the people from above would not let them enter the flow; but people in the stairs from other floors commented that the people from the lower floors disrupted the flow. Even at the stagnation points, the density did not exceed 3 persons/m^2. The authors anticipated a higher density, but attributed the decrease to the fact that it was a drill. It was not indicated if this density included landings and, if so, what area was used for the landings.

Proulx

Proulx (1995) videotaped fire drills in four similar apartment buildings in four different Canadian cities. All of the buildings were 6–7 stories high and were 6–11 years old at the time of the fire drill.

The drills were conducted between 18:45 and 19:30 on weekday evenings in the late summer and early fall of 1993. For all four drills, the weather was sunny and warm. The occupants were given a memo a week in advance notifying them that a drill would be conducted, but the exact day and time of the drill were not indicated.

The pre-evacuation times in the buildings averaged 2.5, 8.4, 9.7, and 3.1 min. The difference between pre-evacuation times appeared to be related to the ability of occupants to hear the alarm according to the author. There were no statistically significant differences in pre-evacuation times based on gender or age.

Once occupants started to evacuate, the average travel time in all four buildings was between 1.1 and 1.3 min. The difference between buildings was not statistically significant. The average speed on the stairs, when it could be measured, ranged from 0.52 to 0.62 m/s. The author did not state if the speeds were along the slope or the horizontal component nor was there any indication for how the travel distance on the landings was calculated. The speeds did include the time when individuals stopped for a rest or to look into hallways. The average time to descend one floor varied from 9.6 to 20.6 s.

Children between the ages of 2–5 years old and the elderly had average speeds of 0.45 and 0.43 m/s respectively. This was slower than the rest of the population, but the elderly occupants tended to leave earlier, thus making their total evacuation times similar to younger adults. People carrying children tended to move at the same average speed as the rest of the population, but they were also more cautious in their movements.

During the evacuations, occupants tended to use stairs that they used on a regular basis even if other stairs were closer to their apartment. The stairs were never crowded during the drill (the density was not reported), but many occupants traveled in groups.

Proulx, Latour, Maclaurin, Pineau, Hoffman, and Laroche

Proulx et al. (1995) recorded evacuations from three high-rise buildings in Canada in 1994. The first two drills (one building in Montreal and the other in Calgary) occurred on weekdays between 18:30 and 19:00 in the summer and fall during sunny and warm conditions. The third building was in Gloucester drill was conducted on a Saturday morning in December between 10:30 and 11:00. The weather for that day was below freezing and snowing.

For all of the drills, cameras were placed in corridors as well as in both stairwells (each building had exactly two). No dimensions were given for the stairs nor were the methods used to calculate travel distance. Speed calculations were based on the total travel distance in the stair and the time required to travel that distance.

Occupants tended to travel in groups and use the nearest stairwell. The groups were described to be moving at the speed of the slowest member and the stairs were described as not being crowded. Occupants were deemed to have a limitation if they were slow and elderly, using a mobility device, carrying things, or assisting other occupants.

The occupants were also grouped according to age, as estimated from the video recordings. The first group was children under 2 years old that had to be carried. The second group was children between 3 and 5 years old that needed assistance on the stairs. The third group was children from ages 6 to 12 years old that did not need assistance, but were typically with an adult. The fourth group was teenagers who may have evacuated without an adult. The fifth group was described as young adults (less than 40 years old). The sixth group was older adults (between 40 and 64 years old). The final group was seniors.

The building in Montreal had 14 floors and 244 apartments. The occupants received a memo 4 days before the drill informing them that a drill would be taking place and that it would be videotaped. It was also noted that the manual normally given to all residents instructed them that, in the case of a fire, they were to wait on their balconies. The drill only lasted 5 min (a previously unknown feature of the system was an automatic silencing of the alarm after 5 min). The average pre-evacuation time of 31 occupants was 90 s. There were no significant differences based on gender, age, or limitation. For descending stairs, the average speed was 1.07 m/s. Fourteen men averaged 1.14 m/s and 15 women averaged 1.00 m/s, but the difference was not significant. The speeds for the different age groups were not statistically significant, but only three of the groups (children 6–12 and the two adult groups) had more than one person. For those three groups, 7 children average 1.30 m/s, 8 young adults averaged 1.00 m/s, and 13 older adults averaged 1.03 m/s. Occupants with limitations traveled at an average speed of 0.88 m/s, also not statistically different, but the authors attributed that in part to only three such individuals being identified. Groups traveled slightly slower than individuals (19 group members averaged 1.00 m/s while 10 individuals averaged 1.18 m/s).

The Calgary building was also 14 floors in height and it had 117 apartments. Once again, occupants received a memo 4 days before the drill. The alarm sounded for 17.5 min. While the elevators were supposed to be recalled, they were not and some occupants used them during their evacuation. Thirty-three occupants had pre-evacuation times that averaged 168 s. Based on gender or age categories with at least five individuals recorded, none of the differences were significant. The nine occupants with limitations required a significantly greater average pre-evacuation time (334 s compared to 106 s). The overall average speed on stairs for 28 occupants was 1.05 m/s. Women moved slightly faster, but the difference was not significant. Nearly all of the movement speeds were from young adults (19 out of 28) with no other group having more than five individuals. Eight occupants with limitations were moving at an average speed of 0.61 m/s (compared to an average of 1.22 m/s for the other 20 occupants).

The Gloucester building was 12 floors tall and had 213 apartments. The drill lasted approximately 20 min. During the drill, 93 occupants required an average of 319 s for pre-evacuation time. For all gender, age, and limitation comparisons (where the category had at least five observations), only seniors had a statistically different time (in this case, they required more time). The average speed for 76 occupants was 0.95 m/s. For descending stairs, men moved statistically faster (1.05 m/s) than women (0.86 m/s). The authors stated that a greater proportion of women being older might have caused this. Based on age, 6 teenagers averaged 1.28 m/s, 30 young adults averaged 1.12 m/s, 21 older adults averaged 0.95 m/s, and 18 seniors averaged 0.56 m/s. All of these speeds (with the exception of teenagers and younger adults) were statistically different. Twenty-one occupants with limitations averaged a statistically slower speed of 0.57 m/s (compared with 1.09 m/s for the 55 occupants without a limitation).

When comparing the three buildings, the average descent speeds were not statistically significant. However the Gloucester building did have a slower speed and the authors attributed this to the colder weather. The authors then combined the populations from the three buildings and found that men (average speed 1.07 m/s) and women (average speed 0.90 m/s) were not statistically significant at their chosen 95% confidence level. However the p-value was 0.06 and that indicates that the speeds were statistically different at the 94% confidence level. Based on age groups, 7 children between 6 and 12 years old averaged 1.30 m/s, 7 teenagers averaged 1.16 m/s, 56 young adults averaged 1.13 m/s, 39 older adults averaged 0.96 m/s, and 20 seniors averaged 0.56 m/s. Seniors were statistically slower than all other groups and the two adult groups were statistically different. Finally, across all three buildings, 32 occupants with limitations averaged a statistically slower 0.61 m/s when compared to occupants without limitations who averaged 1.11 m/s.

Proulx, Kaufman, and Pineau

Proulx et al. (1996) observed evacuation drills in two government office buildings in Canada during the fall of 1995. The weather was overcast with temperatures of 19°C and 13°C. Both buildings were in Ontario (London and Ottawa) and the drills were initiated between 14:00 and 14:15.

Video cameras were located in corridors and in the stairs. The travel distances (in both buildings) changed between floors, but these values (and how they were calculated) were not provided. Movement speeds were based on the total travel distance in the stair and the time required to descend to the exit. Occupants were categorized based on gender and two age groups were also identified (between 20 and 40 years old and between 40 and 65 years old; all occupants were placed into one of the two groups).

For the London drill, the building had seven occupied floors. The drill lasted for approximately 14 min and there were 165 occupants present. Of these occupants, 133 used one of three stairs that was available; the side stair (near the main hall) was used by 66.2% of occupants, the rear stair was used by 22.6% of occupants, and the front stair was used by 11.3% of occupants. Occupants were found to use the more familiar stair rather than the closest stair. For 92 occupants, the average pre-evacuation time was 36 s with women requiring statistically less time (30 s compared to 44 s for men). There was no difference in times based on age. The mean speed for all occupants was 0.78 m/s with the side stair users averaging 0.75 m/s, the rear stair users averaging 0.76 m/s, and the front stair users averaging 0.97 m/s. The rear stair had a smaller width, thus creating a higher density with fewer people. For gender, women traveled at 0.81 m/s and men at 0.72 m/s (the result was statistically significant). The authors noted that this contradicted previous research and they could not explain the reason that this building was different. One possible explanation (not stated in the report) was that women had a shorter pre-evacuation time; they would thus experience a decrease in density, which can lead to faster movement speeds. There were no differences in speed based on age. In a post-evacuation survey, 21.2% of occupants reported believing that it was an actual event and not a drill. The actions of these individuals were not statistically different from occupants that thought it was a drill.

The building in Ottawa had seven full levels and 502 occupants. For this building, the drill lasted approximately 20 min. As with the previous drill, occupants used the familiar stairs with 46.1% using the southeast stair (leading to the main entrance), 26.0% using the northwest stair (leading to the secondary entrance), 14.2% using the southwest stair, and 13.7% using the northeast stair. The average pre-evacuation time for 161 occupants was 63 s. The average speed on the stairs was 0.93 m/s with the southeast stair users having an average speed of 0.82 m/s, northwest stair users having an average speed of 0.92 m/s, southwest stair users having an average speed of 1.1 m/s, and northeast stair users having an average speed of 1.2 m/s. The number of people entering the stair appeared to alter the speed of occupants descending to that level. Men had a statistically faster average speed (0.96 m/s) than women (0.90 m/s). The difference in speeds based on age was not statistically significant. Questionnaires were again provided after the drill and 23.2% reported interpreting the alarm as an actual fire rather than a drill. The only activity that these occupants were more likely to do than other occupants was saving files and turning off their computers. While not stated by the authors, these were not activities that indicate that occupants thinking it was a real event were being more urgent; if anything these activities increased pre-evacuation times.

Shields, Boyce, Silcock, and Dunne

Shields et al. (1997) observed an unannounced drill in an educational building (the main Jordanstown Campus Building of the University of Ulster) during the morning of May 4, 1995. There were 276 persons on the five levels of the building. Two stairs were located within the building with 77% of the population using one stair. The large unequal usage arose from occupants using the more familiar stair rather than the closest stair.

Pre-evacuation time varied depending on where the occupant was located. All office workers had left their room of origin within 144 s. In rooms used for academic purposes, the pre-evacuation time was up to 197 s.

During the evacuation, a wheelchair user was assisted down the stairs. This caused congestion on the stairway, but no one tried to pass the wheelchair party despite there being approximately 40 cm to do so. People behind the wheelchair had an average movement speed of 0.33 m/s and those ahead of the wheelchair had average speeds of 1.1 m/s. The authors noted that the average speeds, based on density, were greater than Pauls had found in his study. Their theory for the cause of this difference was that their building was smaller in height and the difference in speeds could come from fatigue. The authors presented two regression formulas to describe the movement speeds of occupants.

$$s = 1.27 - 0.30 \cdot D \qquad\qquad (A.8)$$
$$s = 1.69 - 0.05 \cdot D \qquad\qquad (A.9)$$

where:

s = speed (m/s)
D = density (persons/m^2)

Equation A.8 was for the population ahead of the wheelchair user and Eq. A.9 was for the population behind the wheelchair user. However the graph of the data in the article showed the line associated with Eq. A.9 having a steeper slope and values that are approximately:

$$s = 1.69 - 0.50 \cdot D \qquad\qquad (A.10)$$

Thus, either the graph or the written equation was in error.

Proulx, Tiller, Kyle, and Creak

Proulx et al. (1999), as part of a study on photoluminescent markings, recorded occupants in four stairs during a building evacuation drill. The building was a 13-story government office building in Ottawa. The drill was conducted at 13:45 on a day with light snow falling.

The emergency evacuation plan called for the fire floor and the floors above and below to be evacuated first with all other occupants waiting on their floor for instruction. The selected fire floor was the tenth floor and instructions to begin evacuating were not given until 6 min after the initial alarm. The building had four stairs located at the corners of the building. One stair had only the photoluminescent markings. A second stair had reduced emergency lighting and photoluminescent markings. The third stair had only reduced emergency lighting. The final stair had normal emergency lighting.

Video cameras were placed at the stair doors for the floors that were to evacuate, at the landing between the fifth and sixth floor in all stairs, and at the exit doors for all four stairs. There were 457 occupants observed during the drill and the average time occupants needed to reach the exit door was 72 s. The drill was completed within 15 min of the initial alarm.

A total of 392 occupants exited from the three floors. In the stair with only the photoluminescent markings, 144 occupants were observed. The average speed of these occupants was 0.57 m/s with a range of speeds from 0.39 to 1.13 m/s. There were 65 occupants in the stair with reduced emergency lighting and photoluminescent markings. The speeds in this stair varied from 0.64 to 1.30 m/s with an average speed of 0.72 m/s. For the stair with only reduced emergency lighting, 82 occupants had speeds from 0.41 to 1.14 m/s and an average speed of 0.70 m/s. In the control stair with normal emergency lighting, there were 101 occupants. Their speeds varied between 0.45 and 0.84 m/s with an average speed of 0.61 m/s. The travel distance in each stair varied, but the authors did not state how they calculated the travel distance or what any of the stair dimensions were.

Based on the control stair having occupants with the second slowest speed (and the order of the average speeds corresponding to the order of the number of occupants), the authors concluded that the difference in speeds was based on the density rather than the photoluminescent markings. Also, counterflow was present in the stair with only the photoluminescent markings.

Density was calculated for the average conditions in each stair during the busiest 3 min of the drill. The densities were 2.05 persons/m^2 (photoluminescent markings stair), 1.00 persons/m^2 (reduced emergency lighting and photoluminescent markings stair), 1.23 persons/m^2 (reduced emergency lighting stair), and 1.30 persons/m^2 (normal emergency lighting stair). How the area used to calculate these values was calculated was not provided.

There were also 30 occupants that left from lower floors despite not being instructed to do so. While the exact speeds were not reported, the authors described these occupants as moving faster since they did not have any density or fatigue issues.

Kratchman

Kratchman (2007) studied two different stairs in a six-story high-rise building during an evacuation drill on a morning in June 2005 at 9:47. A total of 269 occupants were recorded on videotapes located within the stairwells. One stair was 1.44 m wide and the other was 1.54 m wide. The riser height and tread depths were 20.3 and 28.3 cm respectively.

One stair experienced counterflow conditions as firefighters proceeded up the stairs and the other only had unidirectional flow. In the stair without counterflow, the mean speed was 0.80 m/s. For the counterflow case, this value was decreased to 0.70 m/s. These speeds were measured along the slope of the stairs. The travel path on the landing assumed that individuals traveled in straight lines from midpoint to midpoint of each flight of stairs.

In this study, density was measured by counting the number of people over a known area of stair treads in a snapshot every 10 s and assuming that value was constant over the 10 s period. Rather than using the entire buffer zone of 0.3 m recommended by Pauls, the author assumed a buffer zone of 0.13 m accounting for just the handrail projection. Also, rather than using the horizontal area to calculate the density, the author used the area of the sloped surface along the treads. Thus, for the densities to be compared to work done by Pauls, the density in the 1.44 m-wide stair needs to be multiplied by 1.42 and in the 1.54 m-wide stair by 1.40. Also, the reference values that the author used to show that the speeds were less than predicted at a given velocity were for stairs with a smaller riser and tread height. In the

Gwynne and Rosenbaum section later in this appendix (Sect. A.4.10), the reference choice will be shown to be inappropriate based on the assumptions of the reference. For reasons discussed in that later section, the choice might be appropriate in reality, but the author did not present the data in this manner.

When her data is adjusted based on the modifications to density, the data are more in line with the previous findings, but still predict slightly slower speeds. In the speed versus density graphs, nearly all speeds, even at very low densities, were less than the average speed reported earlier in the thesis. The author did not note or explain why occupants in the selected segment that was graphed appeared to be moving slower than the average people across all density values. Because the travel distance used to calculate the speeds on these graphs was different that other places in the thesis, there is probably an error in one of the two the calculation methods. If the actual speeds were in line with the average speed reported earlier in her thesis (and the density adjusted appropriately), then the speeds fall well within the range of the previously cited works.

While it was not statistically significant, occupants in the stair with counterflow sped up as they descended. The author indicated that this might have been caused by a decrease in counterflow at the lower levels. In the stair without counterflow, speeds generally did not increase from the initial speed and there was a decrease in speed when the stair became more crowded at a middle level. During the counterflow, occupants descending moved to the right (inner) exit lane.

Occupants were seen engaging in activities that were not accounted for in the equations (socializing, group behavior, reentry, stopping to let firefighters pass, etc.). Men had a slightly greater average speed than women, but the difference was not statistically significant.

During the evacuation, nearly half of the population was carrying items with women being nearly three times more likely to be carrying an item than men. The primary cause for this discrepancy came from over half of the women carrying a purse or briefcase while less than 3% of men did so. Passing behavior was observed with the faster individual moving to the outer exit lane.

Proulx, Bénichou, Hum, and Restivo

Proulx et al. (2007) used a single fire drill to collect data based on different photo-luminsecent coatings in a 13-story office building in Ottawa, Canada. Approximately 4,000 occupants were in the building during the drill that started at 10:35 am on October 5, 2006 and 1,191 were recorded in the studied stairs. Four of the six windowless stairs in the building were observed during the 6-min duration of the drill.

All four of the stairs had widths of 1.1 m, but no information was reported about the riser height and tread depth. Overhead images of the stair treads were to calculate density. They had similar densities that ranged from 1.56 to 1.60 persons/m^2 during the busiest 5 min of the drill and the maximum density, when the occupants were at a standstill, was 2.30 persons/m^2. The speeds for descent ranged from 0.17

to 1.87 m/s. However how the travel distance was calculated was not provided. The mean speed in three of the four stairs was between 0.57 and 0.66 m/s. While one stairwell had a slower mean speed, 0.40 m/s, this was attributed to individuals with mobility impairments in the stairs rather than the photo-luminescent coatings.

Most occupants required between 1 and 5.5 min to enter the stair. Once in the stair, they tended to stay to the right or form a staggered file at higher densities and to slow down as they descended. The authors attributed this to the merging of additional people from the floors into the stair. Stagnation in the flow was observed near the merging areas. Handrail use also appeared to cause individuals to progress single-file and thus move at the speed of the slowest individual. No passing behavior was observed to overtake individuals using the handrails. At the higher floors, between 70% and 90% of occupants were using handrails. This decreased to 30–60% near the level of discharge. The authors provided two explanations: the lighting was better at the lower level which made the occupants feel safer and thus not hold the handrail or the occupants were starting to button their coats in preparation for going outside.

Hostikka, Paloposki, Rinne, Saari, Korhonen, and Heliövaara

Hostikka et al. (2007), as part of a larger study, observed the evacuation of 281 occupants from a seven-story office building in November 2006. The building occupants were told what day the drill would occur, but not the time of the drill. There were four egress paths usually available, but two of the more commonly used exits were blocked using cold smoke.

Occupants were observed using both video cameras and radio frequency identification (RFID) tags. Only occupants that worked on floors 5 and 6 (82 occupants) were given the RFID tags and these readings were only collected in one of two usable stairs. Only about 60% of the expected RFID tags were read and the lower floors saw higher percentages of RFID tags read than the upper floors.

Queues formed under two different scenarios. First, slower individuals moved to the side to let people from higher floors pass them. Second, one group stopped to let occupants from lower levels enter the stair.

From video images, the flow rates on the lower floors were reported to range from 0.80 to 0.83 persons/m-s. Of the occupants with RFID tags, 44 were recorded on consecutive floors for a total of 97 data points. The movement speeds were calculated based on the difference in times between two RFID readers a known distance apart. The exact means of calculating this distance was not provided and there was no mention of whether the dimensions were effective or total. The density that corresponded to these speeds was approximated by using the flow values from the video data. For densities less than 0.5 persons/m^2, the data was scattered with most observations falling between 0.5 and 1.5 m/s. For densities between 0.5 and 2.5 persons/m^2, the velocity decreased linearly from approximately 0.75 to 0.5 m/s. The median value of a fitted curve was 0.64 m/s. Also, men and women were found to travel at the same speeds.

Peacock, Averill, and Kuligowski

Peacock et al. (2009) collected data from three high-rise buildings that ranged in height from 6 to 18 stories. Typically, 100–300 persons used the stairs during the evacuations. The stairs varied in width from 0.91 to 2.24 m. The riser heights varied from 18.6 to 20.3 cm with tread depths of 25.4–28.3 cm (typographical errors corrected from published paper).

The average speeds in the different buildings varied from 0.40 to 0.83 m/s. While not stated in the publication, these speeds were the slope component. Occupants that started higher in the buildings tended to have slower speeds and the average speed in the 6-story building was the greatest and in the 18-story building it was the least. In the six-story building, the occupants in the stair with firefighter counterflow tended to move more slowly than the stair without counter flow. The 18-story building saw the occupants in the stair with firefighter counterflow move at a faster speed, but that was assumed to be related to the lower level of congestion in that stair.

A multiple regression analysis based on the variables of counterflow, delay time, distance traveled, stair width, density, and gender was conducted. How some of these variables were defined was not specified in the paper, but counterflow was a binary variable for each observation of each individual, delay time and distance traveled were based on the initial appearance on a camera, stair width was the total width, and the density calculation used the entire area of the landing and treads. Gender was found to not be predictive and the other variables accounted for only 13% of the variance.

Blair

Blair (2010) used video recordings of 2,834 occupants from eight stairs in four office buildings (widths varying from 1.12 to 1.38 m, tread depths from 25.4 to 27.9 cm, and riser heights from 17.8 to 19.0 cm) to perform bivariate correlations with the data. Observations were made between adjacent cameras, typically two floors apart, with the travel distance determined using a travel distance that assumed occupants traveled down the middle of the stair with the distance on a landing being twice the stair width. Also, transfer corridors were included and occupants were assumed to have the same travel speed as when on the stairs. Density was calculated over the entire area between floors including the entire landing area. This assumed that the density was uniform over all floors and that the entire area of the stairs and landings was used uniformly. These assumptions led to reported density values that were at least 110% of the values that would have been produced using the methods of previous researchers.

Typically, occupants that started early, or started late, had faster movement speeds than those in the middle. There was a corresponding density pattern that could explain these results. The slope of the density versus speed calculation was similar to that of Eq. A.24, but that curve was found to be an upper bound on the data. In making comparisons to Eq. A.24, seven of the eight stairs had dimensions with k values that were listed. The final stair, from Stair 8 S, the author used 1.08 because of the same 18 cm riser height. However, due to the smaller treads, the k value should have been 1.03 using Pauls's estimates.

Peacock, Hoskins, and Kuligowski

Peacock et al. (2011) used a multiple regression model to predict movement speeds down stairs in nine stairwells in four different office buildings. The stair widths varied from 1.12 to 1.38 m while the tread depths varied from 25.4 to 27.9 cm and the riser heights ranged from 17.8 to 19.0 cm. They recorded 3,031 occupants with each pair of observations of a given occupant resulting in a data point. The main independent variables were stair, gender, carrying objects, exit lane, handrail use, pre-evacuation time, density, and travel distance. Travel distance was determined using the Predtechenskii and Milinskii method. Speeds were calculated from one camera to the next one, typically two floors below it in the building. Transfer corridors were included in the travel distance in some instances. Density was calculated by accounting for the entire landing area and the number of people within that area and it was assumed that this density was constant from the upper camera to the lower one.

Of these variables, only handrail use, travel distance, and pre-evacuation time were not significant. Due to the similarities of the buildings, the finding that stair was significant is likely caused by unaccounted variables and is not a real effect. The model accounted for 21% of the variation within the data. Interaction variables were examined and effects involving density and the stair were significant.

Hoskins

Hoskins (2011) used 13,786 video data points of individuals on 11 stairs in office buildings. The stairs varied in width from 1.05 to 1.38 m. The tread depths and riser heights varied from 25.4 to 27.9 cm and 17.8–20.0 cm, respectively. Typically, observations were recorded on every other floor with the data points being calculated based on the time between observations. When density was used, it was the average value observed as each individual entered the camera view for the upstream and downstream camera. Regression models were used to predict normalized descent times based on gender, flow units, descent rate of the previous flow unit, flow type, exit lane, pre-observation time, number of open treads to previous occupant, and travel distance. Travel distance was calculated based on the occupants traveling in an arc while on the landing and along the slope of the treads. The regression model had an R^2 value of 0.90. When applied to a blind data set, the R^2 value was 0.88. In addition to the regression model, algebraic equations were developed that bounded 97% of the data (both the test data and the blind data) as well as an equation based on a slow moving population. These equations are:emergency conditions.

$$t = 1.93 - 0.51 \cdot D \qquad \text{(A.11)}$$
$$t = 0.30 - 0.35 \cdot D \qquad \text{(A.12)}$$
$$t = 1.93 - 1.67 \cdot D \qquad \text{(A.13)}$$

where:

t = normalized time (s/m)
D = density (persons/m²)

Equation A.11 was for the slowest subpopulation group, Eq. A.12 was a lower bound, and Eq. A.13 was an upper bound.

Additional findings included that gender was significant in descending when the density was light, but not when it was heavy; the pairings between the first and second persons in flow units was not random with respect to gender; fatigue was present in light density conditions; females were more likely to be carrying objects; body size was not significant in predicting movement speed; and occupants stayed to the right during heavy density conditions.

Normal-Use Studies

Similar to fire drills, observations made under normal-use conditions can be used as a basis for predicting movement speeds during emergency evacuations. Most of these studies involve observations made in mass transit or other public facilities. The assumption is that the occupants are primarily concerned with getting to their final destination. Even more so than was the case with the fire drill studies, these studies do not involve fire cues and it is unknown how close actual behavior in an emergency evacuation would be to this data.

National Bureau of Standards

The National Bureau of Standards (1935) observed stairwells under normal, drill, and laboratory conditions; most were normal conditions. Occupant counts were made once the stairwell was filled. At this point, either one individual that could be observed from start to finish or an audible clue (a second researcher that moved with the flow gave a whistle upon entering) started the timing and counting of people that passed the end point. Once the designated person passed the endpoint, the timing and counting ceased.

The report gave the number of people discharged per unit width per unit time and the area per person by dividing the known area by the number of people that were counted. In the discussion here, those numbers are converted back to a velocity using the same assumptions as the authors.

The first stair was at a theater. It was 1.90 m wide with a 19.1 cm riser and 28.6 cm tread. The average discharge rate was 0.601 persons/m-s with a maximum rate of 0.711 persons/m-s. These values corresponded to densities of 1.33 and 1.86 persons/m². For all of the values in this report, the authors were assumed to have used the total width of the stair and to have only used paths along stair treads. If these assumptions are correct, then the discharge rate and density values were 0.714 and 0.844 persons/m-s and 1.58 and 2.21 persons/m² respectively. Thus, the speed was approximately 0.452 m/s for average conditions.

The second stair was 1.52 m wide and located at a transit station. The riser height was 18.4 cm and the tread depth was 30.5 cm. The average and maximum discharge rates were much greater than the theater, 1.04 and 1.53 persons/m-s respectively (1.30 and 1.91 persons/m-s for assumed effective width), and the authors attributed this to greater motivation on the part of the observed people to exit the transit station. The densities were also increased as the average density was 2.20 persons/m² and the maximum density was 2.83 persons/m² (2.74 and 3.53 persons/m² for assumed effective width). The average speed was thus 0.473 m/s.

The third stair was a controlled test at the National Bureau of Standards where people were assembled at the top of the stair and instructed to descend naturally without attempting to run. It was 0.914 m wide and had 17.8 cm risers and 29.2 cm treads. The average discharge rate was 1.70 persons/m-s and the maximum specific flow was 1.86 persons/m-s (2.53 and 2.77 persons/m-s with assumed effective width). The average and maximum densities were 2.63 persons/m² (3.92 persons/m² with assumed effective area). This corresponds to an average speed of 0.646 m/s.

The three evacuations were with different types of occupancies, but there did not appear to be a relationship between movement speed and density. While it occurred in a controlled situation, the highest density was where the movement speed was the greatest.

In addition to the stairs where densities were provided, eight other stairs were presented with just the flow rates. These stairs were from a variety of building types as well as experimental conditions.

The first building with just a flow rate was an office building during a fire drill. The stair was 2.22 m wide with a riser height of 17.8 cm and tread depth of 30.5 cm. The average flow was 1.26 people/m-s (only higher average flow recorded was in the controlled test previously mentioned) and the maximum flow was 1.31 persons/m-s.

A second office building had a stair 2.20 m wide with a riser height of 16.5 cm and a tread depth of 30.5 cm. The average and maximum reported flows were 0.77 and 0.98 persons/m-s. Next, a pair of stairs at two different office buildings had widths of 1.83 and 1.68 m. The riser heights were both 17.8 cm. The tread depth in one building was 27.9 cm and it was 26.7 cm in the other. In both buildings, the average and maximum flows were 0.77 and 0.93 persons/m-s. A fourth office building had a 1.26 m-wide stair with a riser height of 20.3 cm and the tread depth was 28.6 cm. The average and maximum flows were 0.93 and 0.98 persons/m-s. A final office building had a 1.22 m-wide stair with a riser height of 19.0 cm and a tread depth of 29.2 cm. The average reported flow was 0.71 persons/m-s and the maximum flow was 0.77 persons/m-s.

Some observations were also made at transit stations. At one, the stair was 1.83 m with a riser height of 17.8 cm and a tread depth of 29.2 cm. The average flow was 0.93 persons/m-s and the maximum flow was 0.98 persons/m-s. A second station had a 1.52 m-wide stair. The riser height and tread depth were 18.4 and 30.5 cm respectively. The average and maximum flows were 0.98 and 1.09 persons/m-s.

Finally, they reported some flow values found by Illinois Central Railway. The first stair was 1.68 m wide with a riser height and tread depth of 17.8 and 27.8 cm respectively. The average flow was 1.04 persons/m-s. A second stair was 1.22 m wide with a riser height of 17.8 cm and tread depth of 29.2 cm. The average flow on that stair was 1.59 persons/m-s. For two other locations, no stair dimensions were given and the flows were reported as 0.98 and 1.09 persons/m-s.

London Transport Board

The London Transport Board (1958) observed passengers in nine London subway stations. Observations were made by one of three methods. First, to calculate the flow, a researcher recorded the amount of time required for 50 passengers to pass a designated point. The shortest three times were used to calculate the flow. While not stated by the authors, only using the shortest times would serve to give higher estimates of flow than if all values were used. Second, two researchers stood a known distance apart within a crowd. Upon a signal, both researchers started their stopwatches and the one in the upstream position started to move with the flow. The number of passengers to pass the stationary observer were counted and combined with the time and known distance to calculate speed and density. Third, during off-peak times, passengers were recorded walking a known distance in a given time to get the free speeds. The stairs were 1–2 m wide and had between 19 and 23 uninterrupted steps or two sets of 12 steps. No details were given about the tread dimensions or the number of observations. The free-flow speed was 0.98 m/s. At conditions of maximum flow, the speed was 0.67 m/s. The authors did not indicate if this was the horizontal component or the slope component of the speed. At maximum flow conditions, the flow was 1.1 persons/m-s and the density was 1.6 persons/m² (both of these values are assumed to be based on the total width, but the authors do not indicate if this is the case or not).

The authors noted that, for stairs less than 1.22 m, occupants exited proportional to exit lanes 0.53 m wide and, above this threshold, it was proportional to the entire width. This was partially based on a 1.8 m wide stair observed both before and after a handrail was installed down the middle of it. After the handrail was installed, the flow rate dropped to 81% of its original value. The authors attributed this to there previously being three exit lanes being reduced to two exit lanes. If that theory were correct, the flow should have decreased to 67% rather than 81%.

Table A.1 Fruin's recorded descent speeds

Stair dimensions (cm)	Gender	Age (years)	Speed (m/s)
17.8 by 28.6	Male	<30	0.975
		30–50	0.814
		>50	0.670
	Female	<30	0.700
		30–50	0.598
		>50	0.556
15.2 by 30.5	Male	<30	1.10
		30–50	0.957
		>50	0.706
	Female	<30	0.790
		30–50	0.766
		>50	0.664

The authors also found that the flow rate was approximately a constant for most densities studied. As density increased, speed was found to proportionally decrease. Also, the introduction of counterflow was not observed to alter the total flow.

Fruin

Fruin (1971a), as part of his work on pedestrian planning, noted that energy use on stairs was 10–15 times greater for ascent and about one third more for descent than walking on level ground. In addition, the human body tended to sway, resulting in exit lanes of 0.762 m.

For downward travel, movement speeds had a bimodality that the author said indicated that there are two different normal speeds depending on unknown variables. Furthermore, there was a significant difference in speed based on gender and a slightly less decrease in speed based on age.

On an indoor stair with a 17.8 cm riser and 28.6 cm tread, men under 30 years old had an average slope movement speed of 0.975 m/s. For men between the ages of 30–50 years, men over 50 years old, women under 30 years old, women between 30 and 50 years old, and women over 50 years old, the average slope movement speeds were 0.814, 0.670, 0.700, 0.598, and 0.556 m/s. On an outdoor stair with a 15.2 cm riser and 30.5 cm tread, all ages and genders saw their average speed increase generally by a factor between 1.1 and 1.2. For men, the average speeds were 1.10, 0.957, and 0.706 m/s for the age groups from youngest to oldest. The three female age groups, in the same order, had average speeds of 0.790, 0.766, and 0.664 m/s. These values are presented in Table A.1. It appears that the observations were made for only travel on the treads.

Based on energy expenditure, the author recommended that the maximum riser height should be 17.8 cm. The author did not indicate the densities present during these surveys, but he did provide some guidance for the relationships between speed and density.

Fruin divided the different stair density conditions into six levels of service. Level A was 0.54 persons/m^2 or less in which individuals can freely choose their own speed and pass slower individuals. Level B was 0.54–0.72 persons/m^2 in which individuals can freely choose their own speed, but passing slower individuals is difficult. Level C was 0.72–1.08 persons/m^2 in which some individuals are restricted in choosing their own speed and cannot pass slower individuals. Level D was 1.08–1.54 persons/m^2 in which most individuals are restricted in choosing their own speed. Level E was 1.54–2.70 persons/m^2 in which movement speed is severely restricted and intermittent stoppages occur. Level F was 2.70 persons/m^2 or more in which most individuals are stopped; this level of service was not recommended for design under any condition.

A graph, without an accompanying equation, showed an asymptote to the slope speed for densities less than 0.54 persons/m^2 of approximately 0.7 m/s. At higher densities, the speed decreased nonlinearly.

In a separate article Fruin (1971b) used a stadium stair to provide Equation A.14 for movement speeds down stairs.

$$s = 0.650 - 0.097 \cdot D \qquad\qquad (A.14)$$

where:

s = speed (m/s)
D = density (persons/m^2)

The author described four to five treads per person as being the normal spacing on the steps and this was Level of Service C. The maximum density was found to be approximately one person every other tread which equated to being Level of Service F. It was not stated whether this was the horizontal or slope speed or how the density was calculated.

Daly, Mcgrath, and Annesley

Daly et al. (1991) observed passengers descending stairs (and using other components) at eight London Underground stations from November 6 to 14, 1989. Observations were made during both the peak period as well as the inter-peak period. All stairs that were observed had two-way flow conditions and 796 passengers were observed descending the stairs (compared to 496 ascending the stairs). Due to the two-way conditions, the authors made adjustments to the required capacity, but the adjustments were the same as those recommended for level surfaces. Even with these adjustments, the authors noted that the capacities were at or below the expected capacity of 1.14 persons/m-s in all but one observation site. While not stated by the authors, the difference in gait and the more narrow stairs (than many level components) could lead to even minor counterflows having a more significant impact on the flow (Fruin 1971a); this is potentially supported by the finding that the flows were not as great as expected.

The authors did not report the tread dimensions, the number of flights of stairs observed (they did say that, for components including stairs that observations were made at multiple stations and/or on multiple days), or the observed density ranges. They did state that the slant length varied from 5.99 to 9.34 m (thus the assumption was that the speeds reported are along the slope, but this, too, was not explicitly stated) and that the effective width of the stair ranged from 1.13 to 1.80 m after adjusting for boundary layer effects (0.31 m for either a hard edge or handrail).

Based on a curve fit and a base equation form that was used for all types of components studied, the authors proposed that speeds descending stairs (up to the capacity value of 1.14 persons/m-s) could be calculated by Equation A.15.

$$s = \frac{1}{t_0/L + (C/L) \cdot (f/1.14)^{2.7}} \tag{A.15}$$

where:

s = speed (m/s)
L = travel distance (m)
t_0 = free-flow travel time (s)
C = model derived constant (s)
f = specific flow (persons/m-s)

The free speed was found to be 0.67 m/s with a speed at capacity (1.14 persons/m-s) to be 0.56 m/s. Using these values in Eq. A.12, t_0/L is approximately 1.5 and C/L is approximately 0.3.

The authors made the implicit assumption that people change their movement speeds even with very small changes in density from the free-flow conditions. If reality is closer to the assumption made by Fruin, that up to a certain density people can move at their free-flow speed, then the equation will artificially adjust to fit data from both regimes; this would lead to under predictions at lower speeds and over predictions at higher speeds.

Tanaboriboon and Guyano

Tanaboriboon and Guyano (1991) studied four different descending stairs in Bangkok, Thailand. The observations were made under normal use conditions and no consideration was given to the density that corresponded to a given speed. The authors did not indicate if the speed values were for horizontal or slope speeds.

The first stair had a 20 cm riser and 30 cm tread, was 1.20 m wide, and the observation length was 5.00 m. The authors observed 205 individuals with speeds ranging from 0.388 to 0.874 m/s with a mean speed of 0.583 m/s.

The second stair had a smaller riser, 15 cm, and the same tread width, 30 cm, as the first stair. It was 3.00 m wide and the observation length was also 5.00 m. The authors observed 307 individuals with speeds ranging from 0.435 to 0.820 m/s with a mean speed of 0.598 m/s.

The third stair had a 14 cm riser and 30 cm tread, was 1.20 m wide, and the observation length was 5.40 m. The authors observed 140 individuals. The minimum speed observed was 0.440 m/s and the maximum was 0.815 m/s. The mean speed was 0.610 m/s.

The final stair had the smallest riser, 13 cm, and a 30 cm tread like all of the other stairs. It was 1.40 m wide, and the observation length was 4.50 m. The authors observed 215 individuals with speeds ranging from 0.459 to 0.893 m/s with a mean speed of 0.620 m/s.

Lee and Lam

Lee and Lam (2006) used video recordings of stairs in Hong Kong mass transit railway stations on the last two Fridays of January and the first three Fridays of February in 2001. Observations were made using a handheld video camera on a tripod during the morning peak (8:00–10:00), afternoon off-peak (14:00–16:00), and evening peak (17:30–19:30) h. The observation area was delineated using tape on the ground and the precision of the data was 0.04 s.

Their observations were part of a larger study that was intended to provide a better understanding of route selection by occupants in a congested station. The study examined both the unidirectional and bidirectional flow cases for ascending and descending stairs.

The stair had an effective width of 1.94 m with a riser height of 16 cm and a tread depth of 31 cm. Observations were recorded over a length of 5.58 m with the data extraction process being automated. When the stairs were at the peak capacity, the average descending speed ranged from 0.48 to 0.65 m/s. The first number corresponded with a heavy counterflow while the second number was under a unidirectional flow. The authors did not state whether the speed was for the horizontal or slope component of speed.

Within the average flows, some individuals were moving faster as they weaved through the other people. Others went slower than the average flow as well. For descending the stairs, the individual speeds varied from 0.38 to 0.92 m/s and 0.29 to 0.93 m/s for the unidirectional and heavy counter flow cases respectively.

This study is one of the few that has attempted to examine the speeds of individuals within the average flow. No consideration was given as to why individuals chose to go at a speed different than the average nor if there were characteristics that distinguished these individuals from the general population.

Ye, Chen, Yang, and Wu

Ye et al. (2008) made videotape observations of people descending stairs from 8:00 to 10:00 a.m. in one subway station in Shanghai, China from October to November 2006. The stair had 15 cm risers, 30 cm treads, and was 3.05 m wide. Observations were made over a length of 3.35 m as measured along the slope.

The authors calculated density by using 15 s intervals and averaging three points within the interval; rather than calculating the speed of all individuals, the authors only calculated the speeds of three individuals within a given crowd that did not engage in passing behavior. Thus, the values were more representative of the bulk flow rather than individual flow. This resulted in 410 data points. However the authors did not state if they used the horizontal component of density or if they used the slope length to determine the area. They also did not state if they used the effective or total area.

Speeds primarily fell between 0.5 and 1.2 m/s for densities up to 1.7 persons/m^2. Based on their observations, the authors recommended a calculation for the flow per meter:

$$f = 0.996 \cdot D - 0.159 \cdot D^2 \qquad\qquad (A.16)$$

where:

    f = specific flow (persons/m-s)
    D = density (persons/m^2)

While not provided by the authors, with the assumption that the specific flow is the speed multiplied by the density, Eq. A.16 can be used to determine the free-flow speed and maximum density as was done for Eq. A.4. In this case, the free-flow speed would be 0.996 m/s and the maximum density 6.30 persons/m^2.

Post-Incident Studies

In order to better understand how behavior is different in an actual emergency in comparison to drills and normal use, some studies have been conducted that surveyed victims of actual fires. These studies relied on individuals' ability to recall all of their actions during the emergency. In most cases, the authors did not ask respondents to estimate their movement speed. In only a few known studies have the authors attempt to estimate the movement speeds of individuals descending stairs. In these estimates, the exact times were not known and other variables (like density) were either approximated or not collected in combination with the speed values.

Galea and Blake

Galea and Blake (2004) gathered over 250 accounts from the public record of survivors from the World Trade Center collapse. Relying on the public record meant that the survey was not scientific and the accounts tended to be from occupants higher in the building. The information related to 3,291 experiences of 260 occupants.

In Tower 1, the stairs were reported to be crowded below the 44th floor and slow moving between the 44th and 76th floors. In Tower 2, the floors between the 44th and 78th floor were, at least initially, crowded. Below the 44th floor, the flow was fast moving. Occupants typically traveled in groups (90% in Tower 1. 88% in tower 2), but the groups changed size during the evacuation, both gaining and losing members.

Occupants typically reported able-bodied individuals using one exit lane while the other exit lane was reserved for injured occupants (and those assisting them). Water was reported in the stairs and the authors believed that it could have hindered the evacuation. There were some reports of fatigue, typically caused by the nature of footwear worn by some female occupants. Counterflow was also reported as a hindrance to movement down the stairs.

A total of 29 accounts provided enough time cues to estimate descent times, but even this group usually gave an approximate time (or a range of times). Only eight accounts (four in each tower) gave specific initial and final times and did not have extraneous actions. Eight accounts in Tower 1 and three accounts in Tower 2 gave either small ranges of times or had extraneous actions. The final five accounts from both towers were not reliable due to wide time ranges or extraneous actions. In Tower 1, occupants were estimated to travel at 29–33 s/floor (approximately between 0.25 and 0.41 m/s). Occupants in Tower 2 were estimated to travel at a faster speed of 20–29 s/floors (approximately between 0.2 and 0.7 m/s). The m/s calculations were based on occupants traveling down the middle of the stars, making 90° turns, walking half the stair width onto the landing, and then mirroring this path.

Averill, Mileti, Peacock, Kuligowski, Groner, Proulx, Reneke, and Nelson

Averill et al. (2005) conducted 368 telephone interviews with survivors of the World Trade center collapse. In each tower there were two 1.1 m-wide stairs and one 1.4 m-wide stair. Individuals were observed to be altruistic. Some individuals left the stair that they started in due to instructions or deteriorating conditions within the stairs and some women removed their high-heeled shoes. Counterflows were present in some of the stairwells as emergency personnel ascended towards the fire floors. In Tower I, the average movement speed was 0.2 m/s for the entire time in the stairs. This included any rest periods and stopping due to overcrowding in the stairs.

Shields, Boyce, and Mcconnell

Shields et al. (2009) used data collected from interviews from six World Trade Center survivors that had self-identified mobility impairments prior to September 11, 2001. For five of the six occupants, the authors were able to describe the activities and movement speeds (in terms of floors) as they descended.

Participant A, initially on the 64th floor of Tower 1, had recently had knee surgery, had discomfort descending the stairs, and needed to use the handrail for support. While she descended, she allowed others to merge into the flow and deferred to firefighters climbing the stairs. Her speed was also decreased when she encountered water in the stairs. She still managed to descend at the pace of 46 s/floor.

Participant B suffered from hypertension. He was initially located on the 63rd floor of Tower 1 and immediately started herding other people towards the exits. In the stair he experienced some crowding, merging of flows, and deferential treatment towards injured occupants and firefighters ascending the stairs. His rate of descent was approximately 43 s/floor.

Participant C needed an air cast for his injured ankle and was using crutches to move around. During his evacuation from the 54th floor of Tower 1 he did not have a shoe on his injured foot. He entered the stair with 13 other individuals from his floor and they encountered crowding and merging as they descended. Once again, deferential treatment was shown to allow more injured occupants to pass as well as firefighters. He ended up separating from his group and passing people as he descended. As was the case with Participant A, his descent rate was approximately 46 s/floor.

Participant D was initially on the 17th floor of Tower 1. She had multiple mobility impairments and descended immediately with several colleagues. They descended single-file, even though the stair was empty when they entered, so that anyone that attempted to come up the stair could pass them. Around the 10th floor, the person with her offered her part of his shirt to block out the strong smell of jet fuel in the stair. They then had to ascending up three levels due to further passage in that stair being impossible; the second stair had the same result of starting to descend before having to return to a higher floor and change stairs. While descending she required an average of 150 s/floor (including the time needed to ascend and change stairs).

Participant E required canes for moving. She started on the 20th floor of Tower 1 and recruited three colleagues to act as crowd control (so she would not be knocked over) while descending the stairs; one walked behind her, one next to her, and one in front of her. They had to stop every few floors to allow some of her helpers (they had asthma) stop and recover. As they descended, they allowed other people to pass them, but a group formed behind them that refused to pass (one even walked down to her, gave her some water, and then got back in line behind her). They also experienced some crowding as they descended at approximately 75 s/floor.

Galea, Hulse, Day, Siddiqui, and Sharp

Galea et al. (2009) interviewed survivors of the World Trade Center collapse (129 from Tower 1 and 125 from Tower 2). From these accounts, they attempted to approximate the density that was present for each floor, but they did not indicate how the area was calculated. Speeds were calculated along the slope of the stairs, but no direct mention was given about how the travel distance on the landing was measured.

From the accounts of 30 interviewees from Tower 1, estimates of movement speeds down the stairs were approximated. After attempting to eliminate known stops and other identifiable issues, the average adjusted speed was 0.29 m/s. Subgroups of the population, based on body mass index, found no differences in movement speeds. The authors indicated that there was a possibility that the number of stops required by most occupants while descending masked this variable.

From the total survey, not just the interviewees used for the movement speed calculation, the authors found that 86% of interviewees had to stop in the stairs during their descent. Congestion was the most common cause of stopping, followed by allowing others to pass and fatigue; usually a companion that needed to rest.

Compiled Studies

Rather than conducting their own studies, some authors have attempted to develop their own movement speed values for descending stairs based upon the works of other authors or their own perceptions. The sources reported here also include an instance where the authors present the results from another author that was not published in English.

Joint Committee

The Joint Committee (1952) examined different codes and studies from the United Kingdom, Canada, and the United States and looked at the complete situation with respect to the built environment in relation to fire. From these, they made a list of recommendations for all aspects of building design.

For stairs, they recommended that values should be determined from the perspective of 0.53 m-wide units of width. The width used was the entire width and no boundary layer was suggested. Their logic was that, at a basic level, people would walk down stairs as moving files. With a small increase in width, an additional file of people could not be added and there was no data supporting the fact that there would be a linear increase in discharge rate. However the authors did not explain how they determined that 0.53 m was the correct value for an exit lane.

In addition to comments about how people would be in lanes, the authors stated that, from everyday observations, slightly wider stairs seemed to allow greater movement speeds. They could not find this supported in the data of others, so they recommended more testing be undertaken to determine if small increases in width led to higher speeds. However they did add that, if the stair width was only slightly deficient, and the addition of another unit would be unduly expensive, small increases in width could be used to provide adequate means of egress.

The authors found a wide variation in the current data (from 0.60 to 3.4 persons/ m-s). The highest rates were reported in a French study where the subjects (firemen) were told to hurry. They cautioned that these speeds were not expected in most crowds and that, in one test where a subject fell, the speed was then the same as the normal tests.

Based on all of the different codes and studies, they assumed that people in an emergency would behave in an urgent manner and thus recommended that flow rates be 0.94–1.41 persons/m-s (in terms of 0.53 m-wide exit lanes).

For a stair serving only one floor, the authors believed that fatigue would cause occupants to slow down and that the flow should be calculated based on an increase of 8% for every 3.05 m above 6.10 m. They did not state what led to this conclusion.

Galbreath

Galbreath (1969) used the work of the Joint Committee and London Transport Board to develop Equation A.17 for the time required to complete an evacuation.

$$T = \frac{N+n}{r \cdot u} \tag{A.17}$$

where:

T = time
N = persons above the first floor
n = persons per floor or area of stairs divided by 0.3 m²/person (whichever is less)
r = rate of discharge with r_{max} = 2.5 persons/m-s
u = number of 0.56 m exit lanes

The equation could be used with any sets of units since there are no empirical constants in the equation.

The author also estimated that a typical stair would have a travel path of 8.2 m per floor. This consisted of a pair of stair flights that were 1.9 m long and four lengths of the stair width (1.1 m). The area of the stair shaft was said to be 9.2 m². The area was taken by using the length of travel while on the treads.

Using the typical stair dimensions, number of persons/floor (60, 120, 240), and number of levels (15, 20, 30, 40, 50), the author calculated total evacuation times that ranged from 9 to 131 min.

Melinek and Booth

Melinek and Booth (1975), based on the work of Fruin, the London Transport Board, and Togawa made several recommendation for calculating stair movement. They recommend a normal capacity of 1.1 persons/m-s and an unimpeded movement speed of 0.5 m/s. Furthermore, they state that the normal time for an unimpeded crowd to descend one story is 16 s.

Predtechenskii and Milinskii

Predtechenskii and Milinskii (1978) reported that research at the Institute of Architecture of the Russian Academy of Arts (VAKh) found that the speed for descending stairs varied from 0.183 to 0.267 m/s. It is not clear if this was the speed along the slope or the vertical component. The speed was inversely proportional to the density. VAKh recommended using 0.167 m/s for design purposes.

In their list of design parameters (for slope speeds), Predtechenskii and Milinskii recommend using values for every density from 0.01 to 0.92 m^2/m^2 under emergency, normal, and comfortable conditions. The emergency speeds were developed based on the assumption that the 75th percentile speed would be the average under emergency conditions; no measurements under emergency conditions were recorded. For emergency design, the values range from 0.098 to 0.991 m/s. For example, a normal recommended speed of 0.167 m/s corresponds to a density of approximately 0.59 m^2/m^2 with an emergency speed of 0.202 m/s.

The authors also stated, without citing any studies or data, that movement speeds are at a maximum when nearest to a fire and then decrease to normal speeds once far away from the fire.

To calculate the length of travel down stairs, they proposed two equations:

$$L = \frac{2 \cdot L'}{\cos(\alpha)} + 4 \cdot b \qquad (A.18)$$

$$L = L' \cdot \left(\frac{3}{\cos(\alpha)} + 1 \right) + 4 \cdot b \qquad (A.19)$$

where:

L = actual distance traveled (m)
L′ = horizontal projection of path on stairs (m)
α = angle of inclination to the horizontal (degrees)
b = width of stairs and depth of landing (m)

Equation A.18 is valid for one story with two flights of stairs. Equation A.19 is to be used for one story with three flights of stairs.

The authors found that speeds decreased as density increased. These observations were made under normal use and most of the observations were at lighter densities since densities nearing the maximum were very rare. They proposed equation for descending stairs was:

$$s = \left. \begin{array}{c} \left(112 \cdot d^4 - 380 \cdot d^3 + 434 \cdot d^2 - 217 \cdot d + 57\right) \\ \cdot \left(0.775 + 0.44 \cdot e^{(-0.39 \cdot d)} \cdot \sin\left(5.61 \cdot d - 0.224\right)\right) \end{array} \right/ 60 \qquad (A.20)$$

where:

s = speed (m/s)
d = density (m^2/m^2)

Equation A.20 is valid for descending stairs with $d < 0.92$ m^2/m^2. These curves were described to fit well with the experimental data. In emergency conditions, the equation was multiplied by 1.21. According to the authors, if the staircase is steep, the emergency condition multipliers should not be used and might even need to be reduced below normal conditions speeds. The design values that the authors provide do not match this equation and they do not explain how those values were determined.

Similar to Fruin's level of service, the authors proposed different movement behavior based on density. At densities less than 0.05 m^2/m^2, people can overtake slower individuals and engage in passing behavior. Up to 0.15 m^2/m^2, individuals are still engaged in streamline, unidirectional flow, but passing is not readily possible. At higher densities, interactions with other individuals start to determine the movement speed. Up to approximately 0.4 m^2/m^2, individuals can maintain a natural rhythm in their movement. By 0.75 m^2/m^2, contact has become so prevalent that the individuals are moving as a single unit.

The authors recommended that flights of stairs should not have more than 18 steps, the width of the stairs between handrails should be greater than 1.1 m and less than 2.4 m, the slope should be less than 30°, and the treads should be 15 by 30 cm.

Pauls

In an article by Pauls (1984), no new data was presented. Instead the author presented a base equation:

$$\frac{(b - \delta)}{P} = 8.040 \cdot t^{-1.37} \qquad\qquad (A.21)$$

where:

  b = width of the stairwell (m)
  δ = boundary layer (m)
  P = population in the stair (persons)
  t = time (s)

Equation A.21 is an alternate form of Eq. A.1 with the constant value changed slightly.

Based on the author's opinion, not on data (the author specifically stated not to use his values without additional analysis), recommendations were made on how to adjust the $(b - \delta)/P$ value for different riser heights and tread depths. First, 1%, up to a maximum of 10%, was to be subtracted for every 0.5 cm that the tread depth was greater than 28 cm. Second, 1% was to be added for every 0.5 cm that the tread depth was less than 28 cm. Third, 1%, up to a maximum of 10%, was to be subtracted for every 0.5 cm that the riser height was less than 18 cm. Fourth, 1% was to be added for every 0.5 cm that the riser height was greater than 18 cm.

Equation A.4 can thus be rewritten as Equation A.22.

$$s = 1.08 - 0.29 \cdot \left(\frac{1}{\left((b-\delta)/P \right)y} \right)$$

(A.22)

where:

s = speed (m/s)
b = width of the stairwell (m)
∂ = boundary layer (m)
P = population in the stair (persons)
y = length of area (m)

Templer

In his book that dealt with safe stair design, Templer (1992) stated that the average horizontal speed on stairs was 0.45 m/s. How this value was determined was not provided.

Ando

Smith (1995) compiled the work of many previous authors and specifically gave details about work done by Ando in Japan. Ando used extensive (in the words of Smith) observations of passengers in subway stations that were unaware that they were being videotaped. According to Ando, density would reach stagnation when it was 4 persons/m². For movement down stairs, he proposed a linear decrease in speed from approximately 0.9–0.1 m/s as density increases from 0 to 6 persons/m². There was no indication whether this was the horizontal or slope component of speed. While not provided, this would result in approximately Equation A.23.

$$s = 0.9 - 0.13 \cdot D$$

(A.23)

where:

s = speed (m/s)
D = density (persons/m²)

Lord, Meacham, Moore, Fahy, Proulx

Lord et al. (2005) collected information from a variety of sources for different egress components in order to test the predictive capabilities of computer models. The authors did not provide an average value to use, but used the range of data to see the

Table A.2 Crowd movement parameters from Proulx

Condition	Density (persons/m²)	Speed (m/s)	Specific flow (persons/m-s)
Minimum	<0.54	0.76	<0.27
Moderate	1.1	0.61	0.77
Optimum	2.0	0.48	0.98
Crush	3.2	<0.20	<0.66

range of results found in different egress models. They did not mention any attempt to ensure that the speeds were collected under equivalent conditions and measurement methods. In Appendix A of their report, for descending stairs, the authors provide data from a variety of sources for different age groups and weighted all equally. However, for all age groups, more than 71% of the weighted data comes from an article that only looked at people using crosswalks. In Appendix B, different sources are weighted equally for the different age groups. While the references are ones previously cited in this document, several of the results are misapplied. For the children's speed down stairs, data collected on commuters is applied equally to data collected with children. For the 18–29 year old group, once again data collected on all commuters is applied. Also, single observations are weighted equally to averages collected across many studies. Another shortcoming is that some sources referenced other works and both that value and the value from the original are included. In essence, because an author happened to reference a previous work, that work was doubly weighted. This skews the results towards those studies and violates the assumption that the authors made that all studies were equally valid. For occupants 30–50 years old, over 50 years old, and disabled occupants, the same limitations as mentioned for the previous age group were present.

Proulx

Proulx (2008) primarily reported the equations that Pauls (1980) had determined from his original research. The author provided reference values for minimum, moderate, optimum and crush values. Those values are presented in Table A.2.

Under moderate conditions on a 1.22 m-wide stair, each individual would occupy 1–2 treads, have a speed of 0.5 m/s, and descend approximately one floor every 15 s. This is similar to the values found by Kagawa et al. (1985) and level of service C from Fruin (1971a) mentioned previously.

Gwynne and Rosenbaum

Based on the work of Fruin, Pauls, and Predtechenskii and Milinskii, Gwynne and Rosenbaum (2008) proposed calculating stair movement speeds for densities between 0.54 and 3.8 persons/m² (they did not specify what area should be used to calculate the density or what standard conditions would be) by:

$$s = k - a \cdot k \cdot D \qquad (A.24)$$

where:

s = speed along the path of travel (m/s)
k = 1.00 (m/s) for 19.0 cm riser, 25.4 cm tread
   = 1.08 (m/s) for 17.8 cm riser, 27.9 cm tread
   = 1.16 (m/s) for 16.5 cm riser, 30.5 cm tread
   = 1.23 (m/s) for 16.5 cm riser, 33.0 cm tread
a = 0.266 (m²/person)
D = density (persons/m²)

For situations where the density was less than 0.54 persons/m², the speed was assumed to be the speed at 0.54 persons/m². The authors also stated that the specific flow rate could be calculated by Equation A.25.

$$f = s \cdot D \qquad (A.25)$$

where:

f = specific flow (persons/m-s)
s = speed (m/s)
D = density (persons/m²)

Based on the work of Pauls, Fruin, and Predtechenskii and Milinskii, it was possible to determine how the authors developed this equation. The lower limit on Eq. A.24 came directly from the Level of Service concept introduced by Fruin (1971a). It was the boundary between Level A and Level B. Level B was where the very first interactions with other individuals are said to occur. Thus, the free speed was the speed at Level A.

With a 17.8 cm riser height and 27.9 cm tread depth, Eq. A.24 is Eq. A.4. Thus, the equation relied most heavily on the work of Pauls (1980). The highest density Pauls used in developing his model was approximately 2.5 persons/m². The upper bound on Eq. A.24 is 3.8 persons/m² (where the speed becomes 0). Because fitting a curve to data developed the equation, extreme caution should be used when going beyond the data range in which the data was collected. Nowhere did the authors mention this limitation nor did they mention that the equation loses accuracy above 15 stories (as stated by Pauls (1980)). Also, Eq. A.4 was developed using data with densities less than 0.54 persons/m². Because Gwynne and Rosenbaum assumed that speeds less than this limit behaved differently than those above the limit, Eq. A.24 should have been determined with those values being excluded.

The other way that Eq. A.24 differs from Eq. A.4 is the introduction of factors for different stair geometries. As mentioned in Sect. A.1.3, Pauls had a range of stair geometries in his study. While a stair with a riser height of 17.8 cm and tread depth of 27.9 cm was a common stair, it was not the only stair configuration in his study. For proper model development with k-factors, Eq. A.4 should have been recalculated using only a single, common stair dimension. After that, other stair geometries could have been tested to develop the correction factors for the equation.

The development of the k-factors appears to have been determined using the suggestion of Pauls (1984) for four rules for adjusting the $(b-\delta)/P$ ratio. As shown in Eq. A.21 this ratio is inversely related to density. The authors assumed that the horizontal spacing between people was a constant, so the adjustment factor is applied to the default k value of 1.08 m/s. In the absence of data, other reasonable explanations could result in similar k-values.

Ideally, Gwynne and Rosenbaum (2008) would limit the equation to the data range over which it was developed including, the stair geometries, and clearly state that the k-factors are not based on data and/or present information for how k-factors might be determined for unique stair geometries. If that was the case, then other researchers, like Kratchman (2007) mentioned previously, could apply that equation with an understanding of what it means. Under the current formation, and without guidance on how the k-factors should be determined, authors like Kratchman will apply the incorrect k-factor (in her case it should have been approximately 1.03 using the method of Pauls (1984)) to their analysis.

Laboratory Studies

In order to better understand selected variables, some authors have conducted controlled laboratory studies. In these studies, the authors asked the participants to behave in ways such that they could observe specific variables. This often led to conditions that were beyond what the authors expected would be found in real-world situations.

Frantzich – Laboratory and Field

Frantzich (1994) conducted a study that was divided into two parts. The first part consisted of students on a campus stair. The second part looked at a stair in a theater.

For the first part of the study, students were videotaped ascending and descending a 1.3 m wide stair with 27 cm tread depth and 17 cm riser height under controlled conditions. The stair was open on one side with only a handrail, while the other side consisted of a wall and handrail. The subjects knew each other and were instructed to walk at normal speeds rather than to act as if it was an emergency evacuation. All speed measurements were based on the slope of the stair. Individuals descended at 1.0 m/s. For groups, the movement speed varied from 0.82 to 0.91 m/s with densities ranging from 2.2 to 2.5 persons/m². The density was measured by using markers on the stairs, but it was not indicated if the area was effective or total or if the area was calculated along the slope or horizontal.

The second part of the study (a normal use condition) consisted of the general population after a performance in a theater. The subjects were unaware that they were being recorded on videotape and ranged in age from 15 to 70 years. The stair was open on one side with a handrail while the other side had a wall with a handrail. It was 2.25 m wide with a tread depth of 30 cm and a riser height of 15 cm. The movement speed varied from 0.3 to 1.3 m/s, with the speed increasing as the density decreased.

Frantzich – Laboratory

In a study by Frantzich (1996), subjects were students that ranged in age from 20 to 30 years old without any known movement disabilities. The study looked at both ascending and descending movement speeds. Four different stair configurations were used. The first, base case, used a 1.3 m wide stair. The three alternates were: using a 0.9 m wide stair; having two stationary individuals force the flow to move around them, thus increasing the density; and having slower individuals in the stair that were passed. For all configurations, the density was varied from individuals in isolation to large groups. Individuals were recorded on videotapes and the movement speeds were calculated based on the change in shoulder position every 0.25 s. Observations were made along the treads only and the speeds were measured along the slope of the stairs.

The primary stair used in the investigation had a tread depth of 28.0 cm and a riser height of 17.5 cm. The steps were made of brick and the stair was open on one side with a handrail. The other side consisted of a concrete wall and a handrail.

The more narrow stair used in the investigation had a tread depth of 22.5 cm and a riser height of 20.5 cm; this stair had a steeper slope than the wider stair. The steps were made of steel and the stair was open on one side with a handrail. The other side consisted of a gypsum wall and a handrail.

In the trials without any obstructions, the minimal interpersonal spacing was found to be 0.37 m. While not directly stated, it appears that the author was defining this based on the horizontal distance between people rather than along the slope. For the case where the stair width was reduced, the interpersonal spacing was decreased to 0.25 m. Individuals for the case with obstructions were instructed to be as close to the person ahead of them as was possible. Thus, the interpersonal spacing in that trial could be less than would ever be found in an actual evacuation. The movement of individuals near the walls was the same as individuals that were located elsewhere in the stair.

For descending the 1.3 m wide stair without obstructions, the movement speed varied from 0.27 to 1.09 m/s with a mean speed of 0.69 m/s (standard deviation of 0.15 m/s). For the 0.9 m wide stair, the speed varied from 0 to 2.27 m/s with a mean speed of 0.72 m/s and a standard deviation of 0.27 m/s.

The distribution of speeds on either stair size in either direction was approximately normal in this investigation. When comparing the two stairs, the stair width was not significant in determining the movement speed. The author reported that the speeds only decreased slightly as the interpersonal spacing increased.

Based on the results from the 1.3 m wide stair, 0.9 m wide stair, and the interpersonal spacing for the restricted width case, the author proposed that the speed on stairs was a constant (approximately 0.7 m/s for descending) for interpersonal spacing greater than 0.37 m. For interpersonal spacing greater than 0.25 m and less than 0.37 m, the movement speed would increase linearly from 0 m/s to the steady-state value. Interpersonal spacing less than 0.25 m was considered an impossibility and thus the speed would be 0 m/s.

For passing behavior, individuals moved to the side to overtake the slower individual ahead of them.

Boyce, Shields, and Silcock

Boyce et al. (1999) had volunteers with and without disabilities from 5 day centers in the United Kingdom use different egress components. For stairs, occupants first ascended and then descended stairs after a 5-min rest. No information was provided about the number of steps or stair dimensions, but the speed was measured along the slope of the stairs.

Forty-two subjects were able to participate in the stair portion of the experiment. Of these, 8 did not have a disability, 4 needed assistance from another person, and 30 had a disability, but were able to use the stairs without the assistance from another person. The disabled subjects that needed assistance were evenly split between men and women and were all over 75 years old. In the disabled, but not needing additional assistance group, 20 were men and 10 were women with ages from 25 to 85 years old.

The authors noted a few common characteristics about the subjects. First, 94% of the unassisted subjects used the handrails while descending. Second, subjects tended to choose the shortest path while on the stairs, unless they had very little strength in the arm that would be on the handrail side. Third, subjects with locomotion disabilities took up more space descending the stairs than ascending the stairs (this was said to be caused by the ergonomics of stairway movement, but the effect was not quantified).

The eight subjects without disabilities had an average descending speed of 0.70 m/s with a range of speeds from 0.45 to 1.10 m/s. The 30 disabled subjects that were not assisted by another person had a mean speed of 0.33 m/s with a range from 0.11 to 0.70 m/s. Within this group, 19 subjects did not use a mobility aid and had an average speed of 0.36 m/s and their speeds ranged from 0.13 to 0.70 m/s. Nine subjects used a walking stick and had a mean speed of 0.32 m/s and their speeds were between 0.11 and 0.49 m/s. One subject on crutches and one subject that used a rollator had speeds of 0.22 and 0.16 m/s respectively. For the four assisted subjects, the average speed was 0.13 m/s and the range of speeds was 0.11–0.23 m/s.

Wright, Cook, Webber

Wright et al. (2001) conducted a laboratory study in which 18 subjects walked along a smoke-filled egress path under 6 different lighting conditions. There were 7 men and 11 women that ranged in age from 23 to 63 years old (average age 46 years old).

The subjects walked through artificial smoke that varied in mean optical density from 1.1 to 1.2 m^{-1}. The subjects traveled on a landing, down a flight of stairs, and then through two corridors. No dimensions were provided for the stair or if the

Table A.3 Fujiyama and Tyler's movement speeds

Speed type	Stair dimension (cm)	Old	Young
Normal	18.5 by 23.0	0.60	0.76
	17.5 by 25.0	0.71	0.79
	15.7 by 26.7	0.74	0.86
	15.2 by 33.2	0.88	0.96
Fast	18.5 by 23.0	0.80	1.12
	17.5 by 25.0	0.85	1.12
	15.7 by 26.7	0.97	1.25
	15.2 by 33.2	1.11	1.30

speeds were in the horizontal or slope directions. The different lighting conditions were normal lighting, emergency lighting, electroluminescent wayguidance system, miniature incandescent wayguidance system, and two light emitting diode wayguidance systems. The order that occupants proceeded through the different lighting systems was randomized. Under normal and emergency lighting, the average speeds were approximately 0.3 m/s. The other four systems had average speeds that ranged from approximately 0.35 to 0.42 m/s.

Fujiyama and Tyler

Fujiyama and Tyler (2004) had subjects ascend and descend four sets of stairs, with a rest period between flights, inside buildings of University College London. The first stair had 12 steps with a riser height of 18.5 cm and a tread depth of 23.0 cm. The second stair also had 12 steps with 17.5 and 25.0 cm riser height and tread depth, respectively. For the third stair, there were 15 steps with a riser height of 15.7 cm and a tread depth of 26.7 cm. Finally, the fourth stair had nine steps and their dimensions were 15.2 by 33.2 cm for the riser height and tread depth, respectively.

There were 18 subjects, 6 men and 12 women, between 60 and 81 years old and an additional 15 subjects, 7 men, 8 women, between the ages of 25–60 years old. On each set of stairs, the subjects were asked to ascend and descend the stairs twice. For descending the first flight of stairs, occupants were asked to move at their normal pace. For the second set, the occupants were asked to move at their fast pace.

In all cases, the speeds are given in terms of the speed along the slope. The younger group had average, normal speeds that ranged from 0.76 to 0.96 m/s and the older group had comparable speeds of 0.60 to 0.88 m/s. The speeds increased as the slope of the stairs decreased. With the exception of one descending stair, the differences between the two groups were not significant for the normal speeds.

For fast descending of the stairs, the young group had average speeds from 1.12 to 1.30 m/s and the older group averaged 0.80–1.11 m/s. For the set of stairs with the least amount of slope, the difference between the two groups was significant at the 0.05 level. For the set of stairs with the greatest slope, the difference was significant at the 0.001 level. For the two intermediary stairs, the difference was significant at the 0.01 level. All of their calculated speeds are shown in Table A.3.

The authors also collected a limited amount of physiological data on the participants. The height and weight of the subjects were not found to have a high correlation with movement speed on stairs.

References

Averill JD, Mileti DS, Peacock RD, Kuligowski ED, Groner N, Proulx G, Reneke PA, Nelson HE (2005) Occupant behavior, egress, and emergency communication. Federal building and fire safety investigation of the World Trade Center disaster. National Institute of Standards and Technology, NIST NCSTAR 1–7

Blair AJ (2010) The effect of stair width on occupant speed and flow of high rise buildings. MS thesis, University of Maryland, College Park

Boyce KE, Shields TJ, Silcock GWH (1999) Toward the characterization of building occupancies for fire safety engineering: capabilities of disabled people moving horizontally and on an incline. Fire Technol 35(1):51–67

Chertkoff JM, Kushigian RH (1999) Don't panic: the psychology of emergency egress and ingress. Praeger, Westport

Daly PN, McGrath F, Annesley TJ (1991) Pedestrian speed/flow relationships for underground stations. Traffic Eng Control 32(2):75–78

Frantzich H (1994) A model for performance-based design of escape routes. Department of Fire Safety Engineering, Lund Institute of Technology, Lund University

Frantzich H (1996) Study of movement on stairs during evacuation using video analysis techniques. Department of Fire Safety Engineering, Lund Institute of Technology, Lund University

Fruin JJ (1971a) Pedestrian planning and design. Metropolitan Association of Urban Designers and Environmental Planners, New York

Fruin JJ (1971b) Designing for pedestrians: a level of service concept. Trans Res Rec 355:1–15

Fujiyama T, Tyler N (2004) An explicit study on walking speeds of pedestrians on stairs. Presented at 10th international conference on mobility and transport for elderly and disabled people, Hamamatsu

Galbreath M (1969) Time of evacuation by stairs in high buildings. National Research Council of Canada, fire research note No. 8

Galea ER, Blake S (2004) Collection and analysis of human behaviour data appearing in the mass media relating to the evacuation of the World Trade Centre towers of 11 September 2001. Office of the Deputy Prime Minister

Galea ER, Hulse L, Day R, Siddiqui A, Sharp G (2009) The UK WTC 9/11 evacuation study: an overview of the methodologies employed and some analysis relating to fatigue, stair travel speeds and occupant response times. In: Proceedings of the 4th international symposium on human behaviour in fire, Robinson College, Cambridge, pp 27–40

Gwynne SMV, Rosenbaum ER (2008) Employing the hydraulic model in assessing emergency movement. In: DiNenno P (ed) The SFPE handbook of fire protection engineering, 4th edn. National Fire Protection Association, Quincy

Hoskins BL (2011) The effects of interactions and individual characteristics on egress down stairs. PhD dissertation, University of Maryland, College Park

Hostikka S, Paloposki T, Rinne T, Saari J, Korhonen T, Heliövaara S (2007) Evacuation experiments in offices and public buildings. VTT, working papers 85

Joint Committee (1952) Fire grading of buildings part III precautions relating to personal safety. Post-war Building Studies Number 29:22–95

Kagawa M, Kose S, Morishita Y (1985) Movement of people on stairs during fire evacuation drill-Japanese experience in a highrise office building. Fire safety science. In: Proceedings of the 1st international symposium, Gaithersburg, pp 533–540

Keating JP (1982) The myth of panic. Fire J 76:57–61

Khisty CJ (1985) Pedestrian flow characteristics on stairways during disaster evacuation. Trans Res Rec 1047:97–102

Kratchman JA (2007) An investigation on the effects of firefighter counterflow and human behavior in a six-story building evacuation. MS thesis, University of Maryland, College Park

Kuligowski ED, Hoskins BL (2011) Analysis of occupant behavior. In: Peacock RD, Kuligowski ED, Averill JD (eds) Pedestrian and evacuation dynamics, 2010 conference, Springer, New York, pp 685–698

Lee JYS, Lam WHK (2006) Variation of walking speeds on a unidirectional walkway and on a bidirectional stairway. Trans Res Rec 1982:122–131

London Transport Board (1958) Second report of the operational research team on the capacity of footways. London Transport Board research report No. 95

Lord J, Meacham B, Moore A, Fahy R, Proulx G (2005) Guide for evaluating the predictive capabilities of computer egress models. National Institute of Standards and Technology, NIST GCR 06–886

Melinek SJ, Booth S (1975) An analysis of evacuation times and the movement of crowds in buildings. Building Research Establishment, current paper 96/75

National Bureau of Standards (1935) Design and construction of building exits. National Bureau of Standards, miscellaneous publication M151

Pauls JL (1971) Evacuation drill held in the BC hydro building 26 June 1969. National Research Council of Canada, building research report 80

Pauls JL (1980) Building evacuation: research findings and recommendations. In: Cantor D (ed) Fires and human behaviour. Wiley, New York, pp 251–275

Pauls J (1984) The movement of people in buildings and design solutions for means of egress. Fire Technol 20(1):27–47

Pauls JL, Jones BK (1980) Building evacuation: research methods and case studies. In: Cantor D (ed) Fires and human behaviour. Wiley, New York, pp 227–249

Peacock RD, Averill JD, Kuligowski ED (2009) Stairwell evacuations from buildings: what we know we don't know. National Institute of Standards and Technology, NIST technical note 1624

Peacock RD, Hoskins BL, Kuligowski ED (2011) Overall and local movement speeds during fire drill evacuations in buildings up to 31 stories. In: Peacock RD, Kuligowski ED, Averill JD (eds) Pedestrian and evacuation dynamics, 2010 conference, Springer, New York, pp 25–36

Predtechenskii VM, Milinskii AI (1978) Planning for foot traffic flow in buildings. (trans: Sivaramakrishnan MM). Amerind Publishing, New Delhi

Proulx G (1995) Evacuation time and movement in apartment buildings. Fire Saf J 24(3): 229–246

Proulx G (2008) Evacuation timing. In: DiNenno P (ed) The SFPE handbook of fire protection engineering, 4th edn. National Fire Protection Association, Quincy

Proulx G, Latour JC, Maclaurin JW, Pineau J, Hoffman LE, Laroche C (1995) Housing evacuation of mixed ability occupants in highrise buildings. National Research Council of Canada, internal report 706

Proulx G, Kaufman A, Pineau J (1996) Evacuation times and movement in office buildings. National Research Council of Canada, internal report 711

Proulx G, Tiller DK, Kyle BR, Creak J (1999) Assessment of photoluminescent material during office occupant evacuation. National Research Council of Canada, internal report 774

Proulx G, Bénichou N, Hum JK, Restivo KN (2007) Evaluation of the effectiveness of different photoluminescent stairwell installations for the evacuation of office building occupants. National Research Council of Canada, research report 232

Shields TJ, Boyce KE, Silcock GWH, Dunne B (1997) The impact of a wheelchair bound evacuee on the speed and flow of evacuees in a stairway during an uncontrolled unannounced evacuation. J Appl Fire Sci 7(1):29–39

Shields TJ, Boyce KE, McConnell N (2009) The behaviour and evacuation experiences of WTC 9/11 evacuees with self-designated mobility impairments. Fire Saf J 44:881–893

Smith RA (1995) Density, velocity, and flow relationships for closely packed crowds. Saf Sci 18:321–327

Tanaboriboon Y, Guyano JA (1991) Analysis of pedestrian movements in Bangkok. Trans Res Rec 1294:52–56

Templer JA (1992) The staircase: studies of hazards, falls and safer design. The MIT Press, Cambridge

Wright MS, Cook GK, Webber GMB (2001) The effects of smoke on people's walking speeds using overhead lighting and wayguidance provision. Human behavior in fire, In: Proceedings of the 2nd international symposium, MIT Interscience Communications, London

Ye J, Chen X, Yang C, Wu J (2008) Walking behavior and pedestrian flow characteristics for different types of walking facilities. Trans Res Rec 2048:43–51